INTEGRATING MATH IN THE REAL WORLD

THE MATH OF SPORTS

Hope Martin and Susan Guengerich

Portland, Maine

User's Guide
to
Walch Reproducible Books

As part of our general effort to provide educational materials that are as practical and economical as possible, we have designated this publication a "reproducible book." The designation means that purchase of the book includes purchase of the right to limited reproduction of all pages on which this symbol appears:

Here is the basic Walch policy: We grant to individual purchasers of this book the right to make sufficient copies of reproducible pages for use by all students of a single teacher. This permission is limited to a single teacher and does not apply to entire schools or school systems, so institutions purchasing the book should pass the permission on to a single teacher. Copying of the book or its parts for resale is prohibited.

Any questions regarding this policy or requests to purchase further reproduction rights should be addressed to:

Permissions Editor
J. Weston Walch, Publisher
321 Valley Street • P.O. Box 658
Portland, Maine 04104-0658

1 2 3 4 5 6 7 8 9 10

ISBN 0-8251-3920-1

J. Weston Walch, Publisher
P.O. Box 658 • Portland, Maine 04104-0658
www.walch.com

Printed in the United States of America

Contents

Introduction

How often, as math teachers, have we heard students say, "When am I going to use that?" Perhaps this is because much of the mathematics taught in today's classrooms was developed for an industrial society, whose schools were organized and curricula designed to prepare students to work on farms, in factories, and in shops. Today, we live in an information age full of technology that was unimaginable at the beginning of the century. The shop-keeper arithmetic of fifty years ago will not help students succeed in today's world of calculators and high-speed computers.

The National Council of Teachers of Mathematics (NCTM) developed standards, the *Principles and Standards for School Mathematics* (2000), to meet the challenges of the twenty-first century. This document calls for a curriculum that will help all students solve problems and make connections between mathematics and the real world. For mathematics to be relevant to all students, they must see how it relates to their lives outside the classroom.

One way to motivate and interest students is to connect the skills and concepts of school mathematics with real-world applications. In *Integrating Math in the Real World: The Math of Sports,* students can see how topics normally found in the middle-school curriculum can be used to explain and interpret their favorite sports and recreational activities. The activities in this book are open-ended and encourage students to:

- work collaboratively to develop strategies to solve problems;
- make connections between their life experiences and the mathematics classroom; and
- make connections between mathematics and other curricular areas.

Organization of Activities

Each activity in *Integrating Math in the Real World: The Math of Sports* is organized in a similar fashion. Each activity consists of two parts, a teacher page and one or more student pages. Each teacher page includes the following elements:

Areas of Study

Most of the activities in the book relate to more than one strand of mathematics. Often three or four math strands will be addressed in one activity. These strands are listed at the top of each teacher page.

Concepts

The mathematics concepts are listed in the form "Students will . . .". This will help you tie assessment to the lesson objectives. Evaluating the concepts and objectives of the lesson encourages authentic assessment.

Materials

This section includes a brief list of materials, supplies, and special settings the lesson may require.

Procedures

This is a brief description of the lesson and suggestions for the teacher. This section is not meant to be a step-by-step recipe of procedures but an overall guide. Feel free to adapt the procedure to make it more relevant to your classroom and your teaching style.

Solutions

Many of the activities in this book are open-ended, and the answers will depend on your students. Where questions on the student pages have one correct answer, the answer is provided on the teacher page.

Assessment

Suggestions are made in this section concerning assessment strategies. It will suggest observation of students and student groups, and a journal question suggestion, when appropriate.

Extensions

This includes some suggestions you or your students can use to extend the lesson.

Internet Connection

When appropriate, suggested Internet sites are described in this section.

The student pages that follow the teacher page for each activity have been designed to serve as blackline masters and can be copied for student use.

Assessment

When we change the way we teach, we need to examine the way we assess. Paper-and-pencil tests have been used for many years to determine if a student can perform specific skills that have one and only one right answer. An example of a skill is the ability to find the solution to: $x + 5 = 21$. Only one number can be substituted for x to solve this simple equation. This type of evaluation is a natural consequence of a curriculum that is built upon behavioral objectives aligned to discrete skills.

In the *Integrating Math in the Real World* series, problem solving, mathematical communication, and critical thinking skills are emphasized. These are process skills which students work toward attaining on a continual basis. Many times there are no simple right-or-wrong answers and the strategies used are as important to the process as the answer obtained.

How do we evaluate process goals? Multiple strategies are used to assess students' performance. Some are informal, such as observation and questioning, and some are more formal, such as presentations and reports, grading matrices (assessment guides), portfolios, and journals. Suggestions for assessment are provided for each activity and at least one journal question related to the activity is included.

Observation of Students

When students are active and working together, it is essential that the teacher walk around the room to become aware of the progress of the student groups and any problems that might arise. During these times it is possible to assess student understanding in a more formal way. While not every student can be observed each time, it is possible to perform a formal-type assessment at least twice during each grading period for each student. These observations can be shared with both parents and students during parent-teacher conferences.

A form, such as the one on the next page, can be used to make the observations more consistent and simplify the process.

Name of Student ______				
Criteria	**4**	**3**	**2**	**1**
How actively are students participating in group project?				
How well does student appear to understand concept of lesson?				
Is student actively listening to other members of group?				
Is student assuming positive leadership or problem-solving role?				
Comments:				

Using a Rubric for Performance Assessment

Authentic assessment is based upon the performance of the student and should be closely tied to the objectives of the lesson or activity. A rubric can be used to quantify the quality of the work. If the rubric is explained before the activity or project, students become aware of the requirements of the lesson. A grading matrix should be developed in which each of the objectives is examined using a five-point scale.

5 Student shows mastery and extends the concepts of the activity in new and unique ways

4 Student shows mastery of the concepts of the lesson

3 Student shows understanding, but there is a flaw in the presentation or reasoning

2 Student shows some understanding and has attempted completion, but there are some serious flaws in the presentation or reasoning

1 Student makes an attempt, but exhibits no understanding

0 Student makes no attempt

Integrating Math in the Real World: The Math of Sports has been designed to be teacher-friendly. By highlighting the interrelationships between math and sports, both teachers and students are enriched. Students can see the relevance of school skills and concepts to their lives and the interconnectedness of learning.

Connections to the Principles and Standards for School Mathematics

	Problem Solving	Communication	Reasoning and Proof	Connections	Representation	Number and Operation	Algebra	Data Analysis and Probability	Geometry	Measurement
The Official NBA Basketball Court	•	•	•	•		•			•	•
Official Size of Sports Equipment	•	•		•		•	•		•	•
Baseball Statistics: Slugging Percentage and Pitching Statistics	•	•		•	•	•	•	•		
How Fast Did They Go? Some Indy 500 Winners and Their Speeds	•	•	•	•	•	•	•	•		
Jumping Potential	•	•	•	•	•	•	•	•		•
How Fast Can You Throw?	•	•	•	•	•	•	•	•		•
Walking the Indy 500	•	•	•	•	•	•	•	•		•
Design Your Own Game	•	•	•	•	•	•			•	•
Integer Football	•	•	•	•		•	•	•		
Is the Super Bowl Boring?	•	•	•	•	•	•		•		•
At the Olympic Table	•	•	•	•		•		•	•	•
Cost of Television Coverage of the Summer Olympics	•	•	•	•	•	•		•		
World Cup Soccer	•	•	•	•	•	•		•		
Prizes in a Cereal Box	•	•	•	•		•		•		
Women in the Summer Olympics	•	•	•	•	•	•	•	•		
Running the Iditarod	•	•	•	•		•		•	•	•

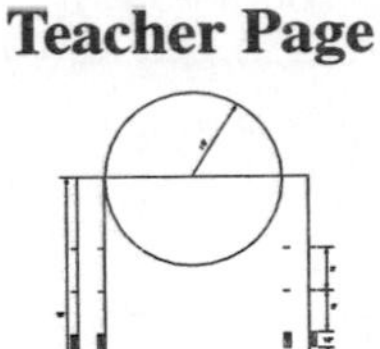

The Official NBA Basketball Court

Areas of Study

Geometry, problem solving, computation

Concepts

Students will:

- use area and perimeter formulas to calculate the size of parts of an official NBA court
- read and understand dimensions on a drawing
- see the connections between geometry and the real world

Materials

- The Official NBA Basketball Court handouts
- calculators
- a list of formulas (optional)

Procedure

The diagram provided includes all the measurements students will need to answer the questions in the lesson. For the most part, students are being asked to use the formulas below to determine the area and perimeter (circumference) of rectangles and circles:

Area of rectangle = $l \times w$

Perimeter of rectangle = $2(l + w)$ or $2l + 2w$

Area of circle = πr^2

Circumference of circle = πd or $2\pi r$

Solutions

1. 4,700 square feet
2. 113.04 – 12.56 = 100.48 square feet
3. area of the out-of-bounds line—48.1 square feet; 2 cans of paint
4. about 14 ft. 3 in.
5. about $56\frac{1}{2}$ in.

Assessment

1. Observation of student
2. Completion of worksheet
3. Journal question:
   - What additional information would you need to calculate the area of the three-point shot?

Extensions

- Have students draw other sports' playing areas to scale and find the area of various parts.

Internet Connection

You can use the Internet to look up your favorite NBA team (or player). You can also find more information about the size of the court or any of the equipment used. An interesting site to start your search is:

http://www.nba.com/

Name ______________________________ Date ______________

The Official NBA Basketball Court

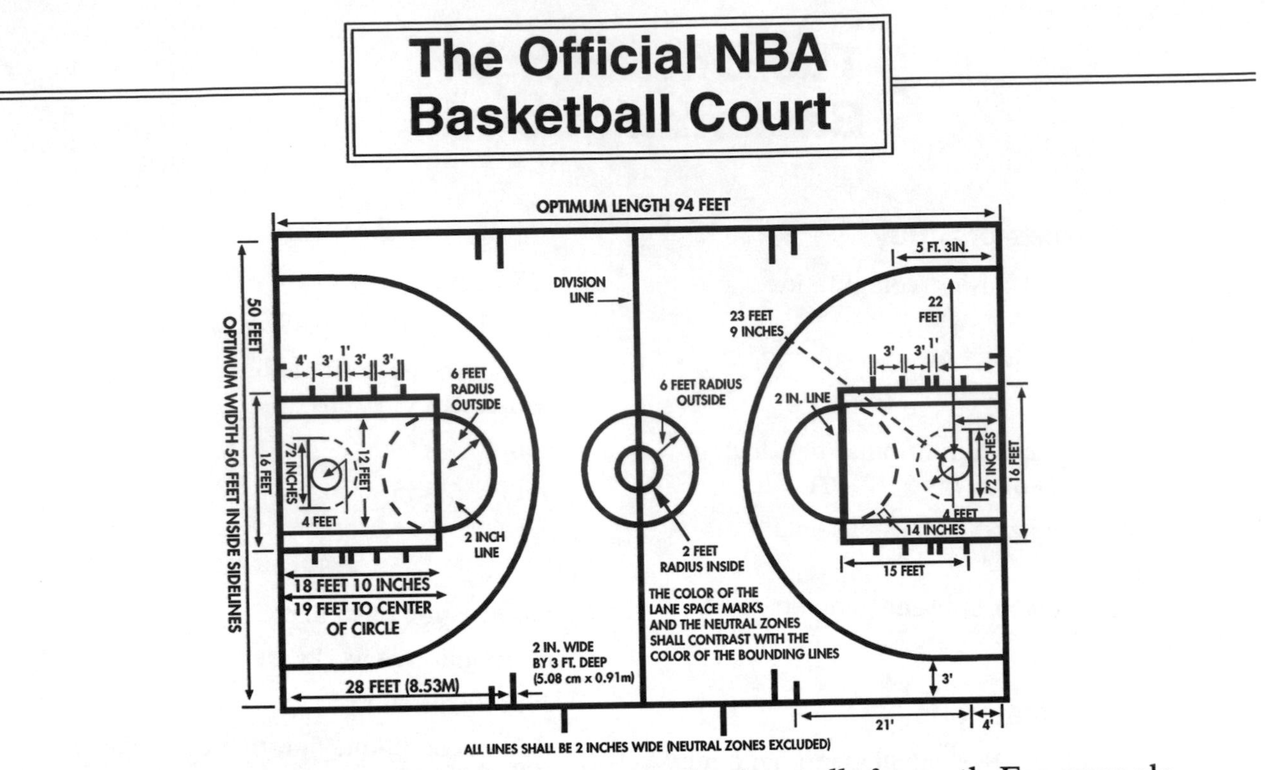

The NBA involves more than just sports; it often calls for math. For example, look at the standards the NBA has set for an official basketball court. Understanding geometry would make it much easier to set up a regulation court.

Use the diagram of an official NBA basketball court above to answer the following questions:

1. Find the area of the entire court (within the out-of-bounds lines).

   ____

2. There are two circles in the center of the court. What is the difference in area between the two?

   ____

3. The out-of-bounds line is 2 inches in width. Find the area of this boundary. If your summer job is to paint this line, and the paint you are using covers 25.5 square feet (ft^2) per can, how many cans of paint will you need?

   ____

4. The distance from the free-throw line to the backboard is 15 feet. About how far is it from the free-throw line to the center of the basket?

   ____

5. What is the circumference of the hoop?

   ____

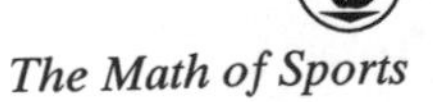

Official Size of Sports Equipment

Areas of Study

Computation, use of formula, radius, diameter, surface area of spheres, volume of spheres, density, analysis of data

Concepts

Students will:

- calculate the diameter given the circumference
- calculate the radius given the diameter
- calculate circumference given the diameter
- calculate the surface area of a sphere given the radius
- calculate the volume of a sphere given the radius
- calculate the density using volume and weight
- order data from least to greatest

Materials

- Official Size of Sports Equipment handouts
- calculators
- various sports equipment, such as tennis ball, golf ball, etc. (optional)

Procedure

Geometric formulas to calculate radius, diameter, circumference, surface and volume will be used extensively in this activity. The size and weight of sports equipment is given, and students will calculate the missing values. Students should be familiar with the following formulas:

Circumference = diameter × π

$d = c \div \pi$

Diameter = $2 \times r$

Radius = $d \div 2$

Surface area of sphere = $4\pi r^2$

Volume of sphere = $4/3\pi r^3$

Density = weight ÷ volume

Calculations should be rounded to the nearest hundredth.

Solutions

Sport	Radius in Inches	Diameter in Inches	Circumference in Inches	Surface Area in Square Inches	Volume in Cubic Inches	Weight	Density
Tennis	1.25	$2\frac{1}{2}$	7.85	19.63	8.18	2 oz.	0.24
Softball	1.93	3.86	$12\frac{1}{8}$	46.78	30.10	6 oz.	0.20
Soccer	4.46	8.92	28	249.84	371.43	16 oz.	0.04
Baseball	1.44	2.87	9	26.04	12.50	5 oz.	0.40
Bowling	4.30	8.60	27	232.23	332.87	16 lbs.	0.77
Golf	0.84	1.680	5.28	8.86	2.48	1.620 oz.	0.65
Handball	0.94	$1\frac{7}{8}$	5.89	11.10	3.48	2.3 oz.	0.66
Lacrosse	1.12	2.23	7	15.76	5.88	5 oz.	0.85
Basketball	4.54	9.08	28.5	258.88	391.77	19 oz.	0.05

Assessment

1. Observation of students
2. Completed worksheet
3. Classroom discussion of various sports equipment
4. Journal questions:
   - Using the data on the chart, write a question that asks a classmate to compare three different spheres.
   - What difference would it make if the official sizes of sports equipment were made smaller, larger, heavier, or lighter?

Extensions

- Use the given data to make a bar graph comparing the various sizes of sport equipment.
- Investigate the size of the basketball used in the NBA. What differences do you find between the NBA basketball and the basketballs used by the NCAA and the WNBA?
- Use the library or the Internet to find the official sizes of other sport equipment.
- Use a computer spreadsheet program to calculate the data.

Name ________________________________ Date ________________

Official Size of Sports Equipment

Sports organizations regulate the size and weight of equipment used in sports. This is to make sure that all players have an equal and fair chance to win. The table below shows the official sizes and weights of various balls. The rules sometimes specify the diameter of a ball, sometimes the circumference. Use the information supplied in the chart and the formulas at the top of each column to calculate the following for each ball: radius, diameter, circumference, surface area, volume, and density. Use 3.14 for the value of pi. Round the answers to the nearest hundredth.

Sport	Radius $2r = d$	Diameter $d = 2r$	Circumference $c = d\pi$	Surface Area $SA = 4\pi r^2$	Volume $V = 4/3\pi r^3$	Weight	Density weight ÷ volume
Tennis		$2\frac{1}{2}$"				2 oz	
Softball			$12\frac{1}{8}$"			6 oz	
Soccer			28"			16 oz	
Baseball			9"			5 oz	
Bowling			27"			16 lbs	
Golf	1.680"					1.620 oz	
Handball	$1\frac{7}{8}$"					2.3 oz	
Lacrosse			7"			5 oz	
Basketball NCAA; WNBA			28.5"			19 oz	

Baseball Statistics: Slugging Percentage and Pitching Statistics

Areas of Study

Problem solving, substitution into a formula, computation, percentages, mathematical connections, reading data from a table

Concepts

Students will:

- read data from a table and use it to calculate well-known baseball statistics
- substitute data into formulas
- calculate percentages using ratios

Materials

- Baseball Statistics handouts
- calculators

Procedures

Slugging Percentage: This handout compares the commonly used statistic, "Batting Average," with a more descriptive, but less well-known, one, "Slugging Percentage." The slugging percentage is a weighted ratio based upon the number of bases the batter gets to with each hit. For example, the number of home runs is multiplied by 4 because the batter crosses four bases, the number of triples is multiplied by 3, and so forth. Students should be asked to predict which player has the largest slugging percentage before beginning their calculations.

When calculating the number of singles (or one-base hits), it is necessary to perform some calculations. For example: in 12,364 times at bat, Hank Aaron had 3,771 hits during his career; 755 were home runs (H), 98 were triples (3B), 624 were doubles (2B). The table does not indicate the number of singles. To find this number, students must add 755 + 98 + 624 to find that Hank Aaron had 1,477 hits that were not singles. By subtracting from his total of 3,771 hits, we find that 2,294 of his hits were singles (1B). Now we have all the information we need to find Hank Aaron's slugging percentage:

$$\frac{4(755) + 3(98) + 2(624) + 1(2{,}294)}{12{,}364} = .555$$

Students can also be asked to order the batting averages (from least to greatest) using players' names and then do the same with the slugging percentages.

Pitching Statistics: The tasks for finding the earned run average (E.R.A.) and winning percentage of pitchers are similar to slugging percentage, in that students have a formula to work with and must substitute the appropriate information for the variables.

The purpose of the E.R.A. is to place all pitchers on an even playing field. Some pitchers will start a game and pitch six or seven innings while others will enter a game toward the very end and pitch to a limited number of batters—sometimes only one! It would be unfair to judge all pitchers merely by the number of runs they gave up. So the formula for finding the E.R.A. includes multiplying the number of earned runs the pitcher has given up by 9 and dividing by the number of innings they've pitched. By following these steps, we are calculating the number of runs the pitcher might give up in a 9-inning game. Have students predict who they believe has the smallest E.R.A. and the highest winning percentage.

Solutions

Slugging Percentage

Hank Aaron—.555
Ernie Banks—.500
Joe Dimaggio—.579
Mickey Mantle—.557
Roger Maris—.476
Jackie Robinson—.474
Babe Ruth—.690

Pitching Statistics	E.R.A.	Winning %
Dwight Gooden	1.53	86%
Greg Maddux	1.56	73%
Roger Clemens	2.48	86%
Sandy Koufax	1.74	79%
Fergie Jenkins	2.77	65%

Assessment

1. Observation of student
2. Grading matrix
3. Journal questions:
   - What is the only time that the batting average and slugging percentage will be the same number? Is it ever possible to have a larger batting average than slugging percentage? Explain your answer.
   - If a pitcher has a winning percentage of 75%, what possible ratio of games won to games won and lost might he have? (There is more than one right answer.)

Extensions

Students can explore other formulas that are used to help managers and owners rank players. Most of these can be found on Internet sites that emphasize statistics.

Internet Connection

One of the more interesting Internet sites for major league baseball information is:
http://www.majorleaguebaseball.com/

Name ______________________ Date ______________

Baseball Statistics: Slugging Percentage

The statistic most often used to describe a batter's ability to get hits is the batting average. This is the ratio of the number of hits the batter has gotten to the total number of times he/she has been up at bat. The batting average would be calculated in the following way:

$$\textbf{Batting Average} = \frac{\textbf{Hits}}{\textbf{At Bats}}$$

This average does not tell us how powerful a batter is: does the batter get mostly singles, or is this person a home-run slugging power? There is a statistic that is a weighted average, giving more points for a home run than for a single. This statistic is called the **slugging percentage**, and is determined as follows:

$$\textbf{Slugging Percentage} = \frac{\textbf{4(HR) + 3(3B) + 2(2B) + 1B}}{\textbf{At Bats}}$$

The table below shows the statistics for seven of the most famous all-stars baseball has ever known. Their lifetime batting average is given, but their slugging percentage is not. Use the formula above to discover who is the most powerful hitter amongst these famous players from the past.

Batter	Avg.	AB	H	2B	3B	HR	Slugging Percentage
Hank Aaron	.305	12,364	3,771	624	98	755	
Ernie Banks	.274	9,421	2,583	407	90	512	
Joe Dimaggio	.325	6,821	2,214	389	131	361	
Mickey Mantle	.298	8,102	2,415	344	72	536	
Roger Maris	.260	5,101	1,325	195	42	275	
Jackie Robinson	.311	4,877	1,518	273	54	137	
Babe Ruth	.342	8,399	2,873	506	136	714	

Key: Avg.—batting average; AB—at bats; H—hits; 2B—doubles (2-base hits); 3B—triples (3-base hits); HR—home runs (4-base hits)

Find your favorite baseball player's statistics in the newspaper or on the Internet. Use the statistics to find the player's slugging percentage and compare that to his batting average. Which of these statistics is greater? How would you explain the difference?

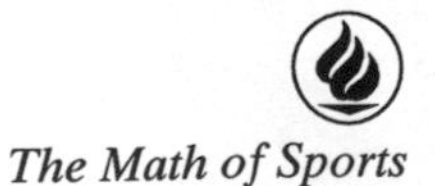

Name ______________________________ Date ______________

Baseball Statistics: Pitching Statistics

Major league pitchers are judged on a few different ratios. Some are based on the number of earned runs batters get while they are on the mound, and some are based on the number of games won while they were pitching.

The first—and best known—statistic is the earned run average, or E.R.A. This is the ratio of the number of earned runs given up during the number of innings pitched. An earned run is a run that is not the result of an error on the part of one of the players.

The formula for the E.R.A. is:

$$\frac{\textbf{Earned runs} \times 9}{\textbf{Innings pitched}}$$

The Cy Young Award is given each year to the best pitcher(s) in major league baseball. Below are the records of five Cy Young Award winners. Use the formula above to calculate each pitcher's E.R.A. for the year he won this prestigious award. Record this number in the E.R.A. column.

Name of Player	Year	Wins	Losses	Innings Pitched	Earned Runs	E.R.A.	Winning %
Dwight Gooden	1985	24	4	$276\frac{2}{3}$	47		
Greg Maddux	1994	16	6	202	35		
Roger Clemens	1986	24	4	254	70		
Sandy Koufax	1964	19	5	223	43		
Fergie Jenkins	1971	24	13	325	100		

Now use the information of the table to find the Winning Percentage for each pitcher. The Winning Percentage is the ratio between the number of wins and the innings pitched:

$$\textbf{Winning Percentage} = \frac{\textbf{Games won}}{\textbf{Games won and lost}}$$

Calculate the Winning Percentage for each pitcher and write your results on the table above.

How Fast Did They Go? Some Indy 500 Winners and Their Speeds

Areas of Study

Statistics, reading data from a table, box-and-whisker plots, quartiles

Concepts

Students will:

- order a set of data from least to greatest
- use the data to find the median, upper quartile, and lower quartile
- design a box-and-whisker plot
- analyze the graph to make predictions

Materials

- How Fast Did They Go? handouts
- calculators

Procedures

The table on the student pages lists selected Indy 500 races from 1911 through 1998. Students might expect that the speeds would go up each year because of progress in designing engines, but the table shows this is not true. One reason for this might be weather conditions. Another might be the yellow flag raised when there is an accident; when the flag is raised, every car must stay in the same order and reduce speed. If there are many yellow flags, this would affect the overall average speed of the race.

Students will first order the data from least to greatest and find the range of speeds. The slowest speed is 74.6 mph and the fastest 176.5 mph; the range is 101.9 mph.

The line graph has 27 units; the range of the data is about 102. $102 \div 27 \cong 4$. If we start numbering at 74 and increase by 4, we will be able to graph all the data.

The median of the data is 128.5; the lower quartile is 96.6; the upper quartile is 148. The range is from 74.6 to 176.5. Remember, each of the quartiles contains the same number of speeds—if the size of the section is smaller, the data are closer together; if the size of the section is larger, the data are further apart.

Students are asked if each of the quartiles is about the same size and they generally are—the second quartile is a little larger (meaning that the scores are a little more spread out in this section).

Solutions

Range: 101.9

Median: 128.5

Student plots should look like this:

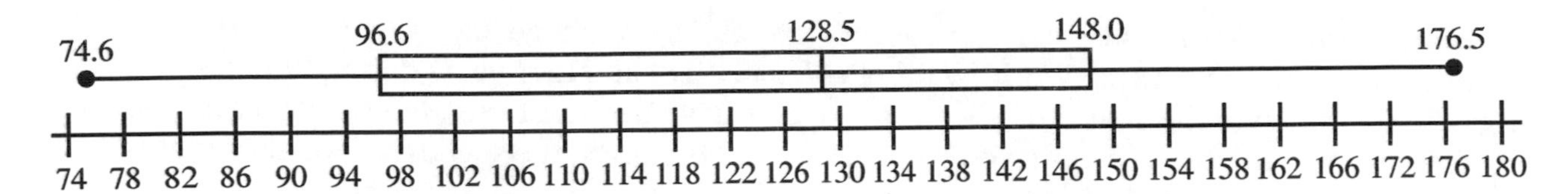

Assessment

1. Observation of student
2. Quality of box-and-whisker plot
3. Journal question:
   - You have a box-and-whisker plot that looks like this:

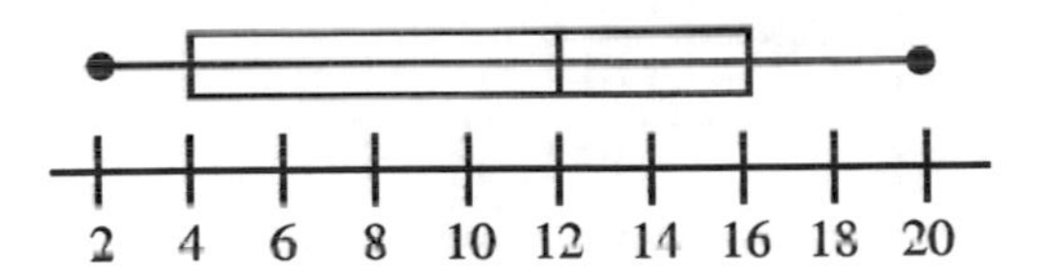

Describe the data, what each quartile stands for, what the range of the data is, and what possible scores might be if there were 17 pieces of data.

Extensions

- Students can measure their feet to the nearest millimeter and make a box-and-whisker plot of the data.

Internet Connection

Interesting sites on the Internet:

http://www.indy500.com/

http://www.sportsline.com/u/racing/auto/indy/

Name ______________________ Date ______________

How Fast Did They Go? Some Indy 500 Winners and Their Speeds

The Indy 500 race has been run over Memorial Day weekend each year since 1911, except for two years during World War I (1917 and 1918) and four years during World War II (1942–45). Over the years, the speeds of the cars have become faster, but the data are very hard to analyze. Depending on the condition of the track, the speeds do not always go up. Look at the table below. It lists some of the winners, the year the race was run, and the average speed of the winning car. What do you notice about the times? Speeds have been rounded to the nearest $\frac{1}{10}$ of a mile/hour.

Year	Winner	Average Speed in miles per hour
1998	Eddie Cheever, Jr.	145.2
1996	Buddy Lazier	148.0
1991	Rick Mears	176.5
1986	Bobby Rahal	170.7
1981	Bobby Unser	139.1
1976	Johnny Rutherford	148.7
1971	Al Unser, Sr.	157.7
1966	Graham Hill	144.3
1961	A.J. Foyt	139.1
1956	Pat Flaherty	128.5
1951	Lee Wallard	126.2
1946	George Robson	114.8
1941	Floyd Davis/Mauri Rose	115.1
1936	Louis Meyer	109.1
1931	Louis Schneider	96.6
1926	Frank Lockhart	95.9
1921	Tommy Milton	89.6
1916	Dario Resta	84.0
1911	Ray Harroun	74.6

Use this space to order the speeds from least to greatest: ______________________

____

____

______________________________________ *(continued)*

Name ______________________ Date ______________

How Fast Did They Go? Some Indy 500 Winners and Their Speeds *(continued)*

What is the range of the data? ______________________

What does this range mean? ______________________

Now that the data are ordered, we can use them to make a graph called a box-and-whisker plot. For this type of graph, we divide the data into four sections; each section must contain 25% of the data.

Directions:

1. Find the **median** (or middle score) of your ordered set of data. Circle it. This speed divides the speeds into the upper and lower 50%.
2. Find the median of the lower half of the data. Circle it. This speed divides the lower half of the data into two **quartiles**. Each quartile contains 25% of the data.
3. Find the median of the upper half of the data. Circle it. This speed divides the upper half of the data into two **quartiles**. Again, each quartile contains 25% of the data.
4. Use the line below to form the box-and-whisker plot. First, mark off a scale on the line that allows for the range of the data. What was the fastest speed? the slowest? The scale must be of even intervals and allow for all the scores between these two extremes. Ask yourself the following questions:
   (a) How many units do I have on the line? ________
   (b) What is the range of the data? ________
   (c) How far apart can my units be if I start at 74 and end at about 177? ________
5. Mark off the units on the line.
6. Place dots at the highest and lowest speeds.
7. Place lines about 1 cm long at the lower quartile, the median, and the upper quartile. Connect the lines; this is the box. Draw a line from the lowest to the highest point; this is the whisker.

Do all of the quartiles appear to be the same size? Could you use this graph to predict approximately what the winning speed might be next year? Explain your answer.

____

____

____

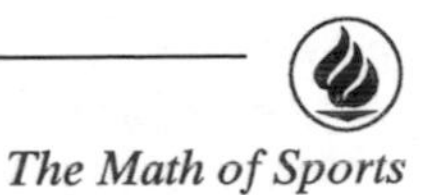

Jumping Potential

Areas of Study

Measurement, computation, percentages, percent increase, data collection, analysis of data, prediction

Concepts

Students will:

- measure their height and jumping height in centimeters
- calculate the difference of their height and jumping height
- calculate their jumping potential (percent increase of their height to jumping height)
- make a scatter plot with height and jumping height
- make a scatter plot with height and jumping potential (% increase)

Materials

- area where students can measure heights and jump height
- Jumping Potential handouts
- calculators
- meter sticks or tape measures

Procedure

Students begin by working in pairs to measure their height in centimeters. Meter sticks or tapes secured to a wall work well for this. Next, students should also measure their jumping height (the number of centimeters they can reach by jumping with one hand raised over the head). For this part of the activity, use of the gym, and the cooperation of the coach, would be helpful. Students should record this data.

Students then work individually to find the difference of their height and jumping height. To calculate their jumping potential, students will calculate the percent increase of height to jumping height. The following formula is used:

$$\frac{\text{jumping potential}}{100} = \frac{\text{difference (jumping height} - \text{height)}}{\text{height}}$$

Students will collect class data and make a scatter plot using height and jumping height. The trend of the data should be the taller the students, the higher the jumping height. The second scatter plot will be constructed using height and jumping potential. The trend may not be evident in this graph because many shorter students may have greater jumping potential.

Assessment

1. Observation of student's work and measurements
2. Completed worksheet and scatter plots
3. Classroom participation while taking measurements
4. Journal questions:
   - When calculating percent increase or jumping potential, why was the student height considered the original measurement?
   - What three students in your class should receive an Outstanding Performance Award? Why?

Extensions

- Use the given data to find the range, mean, mode, and median height in your class. Compare your class to other classes in your building.
- With the help of the coach, determine the jumping potential of your basketball team.

Name ______________________________ Date ______________

Jumping Potential

Your **jumping potential** is a mathematical calculation that tells what percent increase of your height you can jump. To make this calculation, you will need to know your height and how high you can jump.

1. With the help of a classmate, measure your height in centimeters.

   My height is ______________ centimeters.

2. With the help of a classmate, measure your jumping height. This is the height you can reach by jumping with your arms raised over your head. Your teacher will set up an area to measure this. Be sure to follow instructions.

   My jumping height is ______________ centimeters.

3. Find the difference of your jumping height and your height by using subtraction.

   The difference of my jump and my height is ______________ centimeters.

4. Find your jumping potential by calculating the percent increase using this ratio.

$$\frac{\textbf{jumping potential}}{\textbf{100}} = \frac{\textbf{difference (jumping height – height)}}{\textbf{height}}$$

5. Collect the data from your class and record on the class data sheet.

6. Determine the difference and the jumping potential for all the students in your class.

7. Make a scatter plot using the height and jump height of your classmates. Is there a pattern? Do the tallest people have the tallest jump? Describe any patterns you may see. Can you use this scatter plot to make predictions about the jumping height of other students?

8. Make another scatter plot using height and jumping potential. Is there a pattern? Do the tallest people have the highest jumping potential? Describe any patterns you may see. Can you use this scatter plot to make predictions about the jumping potential of other students?

(continued)

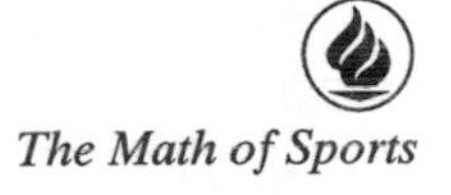

Name ____________________ Date ____________________

Jumping Potential *(continued)*

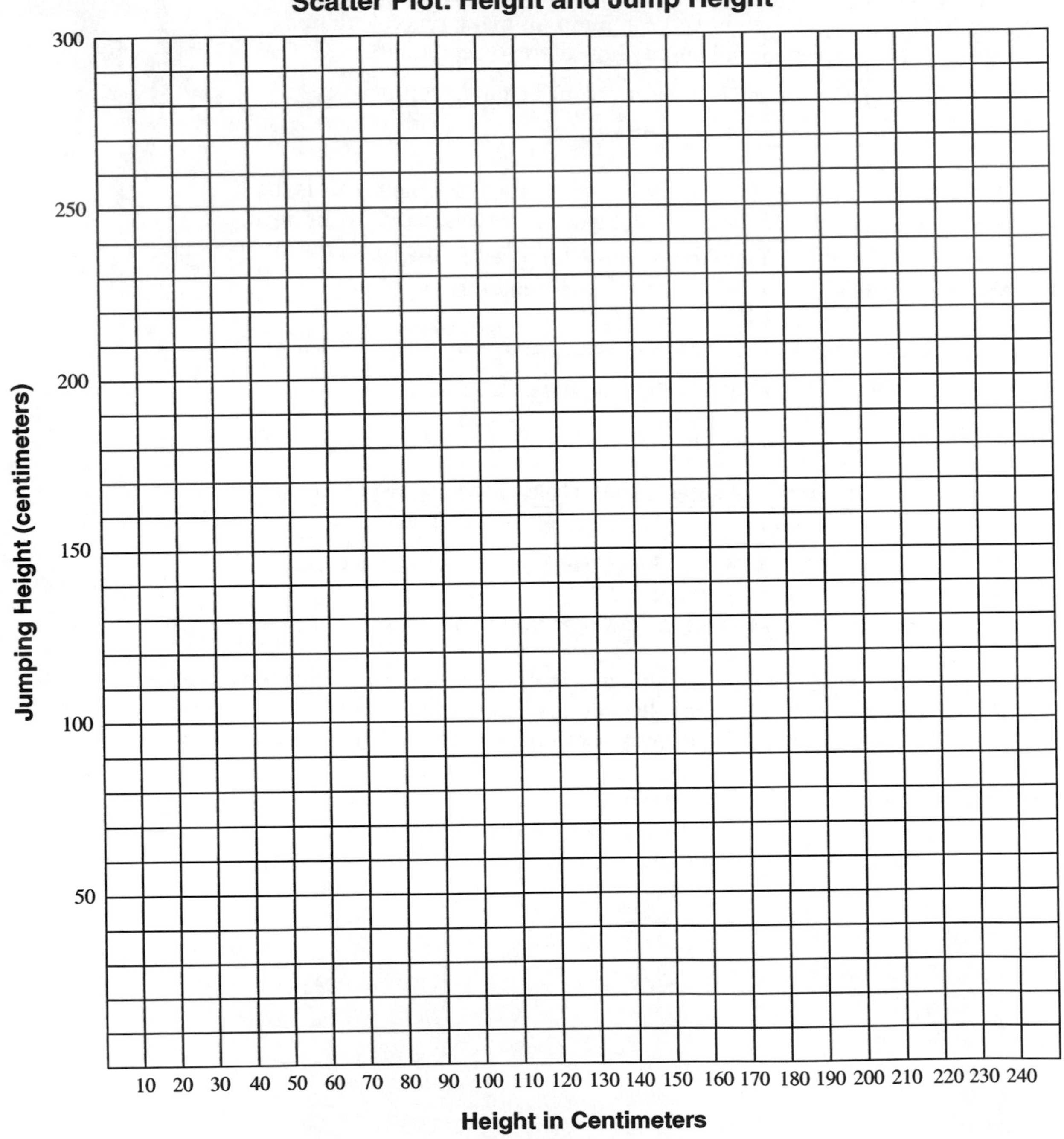

(continued)

Name ______________________________ Date ______________

Jumping Potential *(continued)*

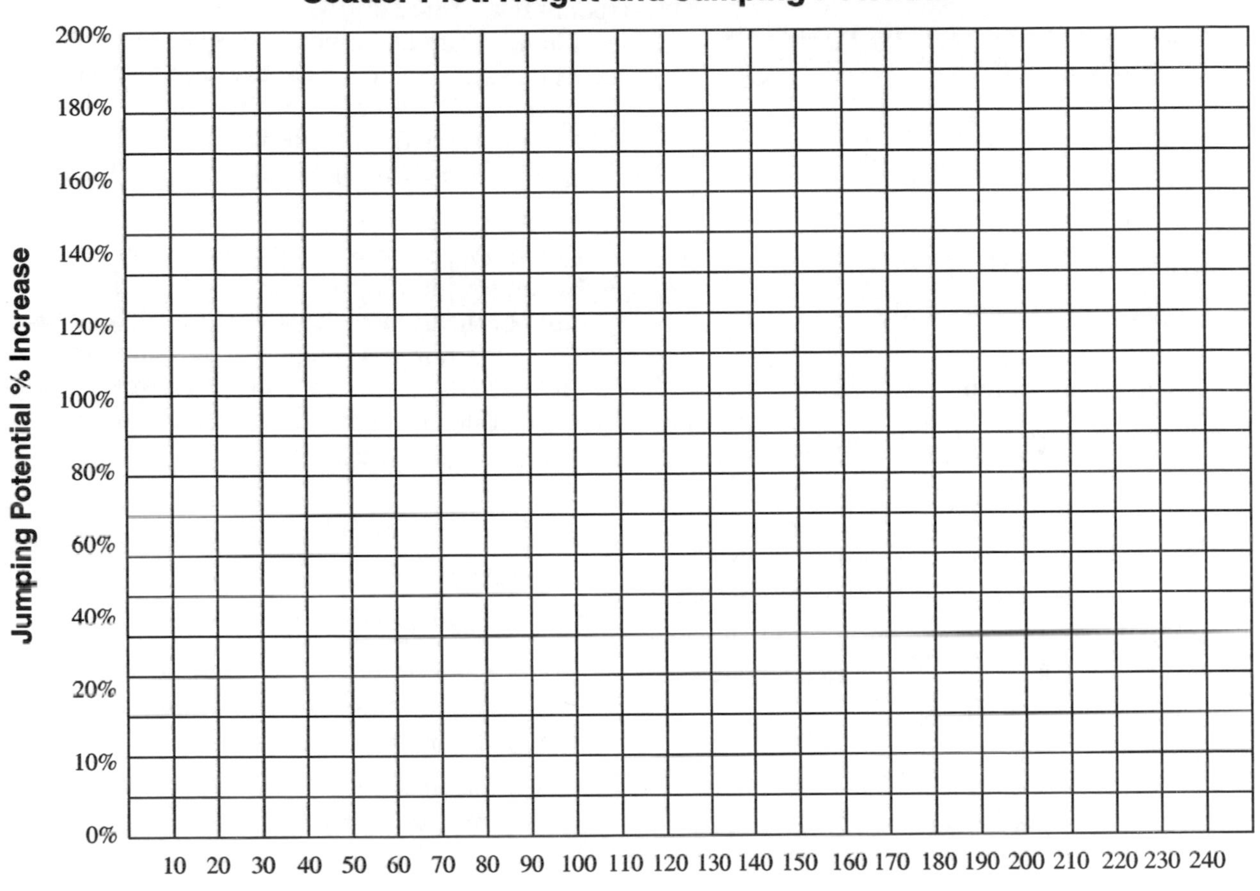

How Fast Can You Throw?

Areas of Study

Problem solving, substituting for variables in formulas, computation, statistics, measurement, collaborative learning

Concepts

Students will:

- use a stopwatch to find time
- substitute for variables in the $r \times t = d$ formula
- work in groups of four to collect data
- calculate the average rate for their class

Materials

- overhead transparency of Class Data sheet
- How Fast Can You Throw? handout (for each group of four)
- digital stopwatch that records time in tenths of a second (for each group)
- calculator (for each group)

Procedure

Divide class into groups of four. Review the distance formula with students and demonstrate the transformations involved with changing it from: $\mathbf{d = r \bullet t}$ to $\mathbf{{}^{d}/_{t} = r}$.

Each group will need a stopwatch to record the approximate time it takes for the ball to reach its destination.

Measure a predetermined distance from a wall and mark that distance off. Be sure that all students can throw the marked-off distance. Each student takes three throws and the average time is used in the Class Data sheet. Let's use these statistics as one student's data: the distance marked off (d) is 10 feet. The average time for the three trials is 1.8 seconds (t). By substituting into the formula: ${}^{10}/_{1.8} = r$. The rate for this student is $\cong$ 5.6 ft/sec.

As each student takes a turn throwing, one group member records the data, one keeps track of the time (with the stopwatch) and the third member of the group calculates the average time. When each group has completed their measurements, students can record their mean time on the Class Data sheet. Together, calculate the average rate of speed for the class.

Assessment

1. Observation of student
2. Grading matrix
3. Journal question:

   Explain how the distance formula would look to find the:

   (a) distance we traveled (if we know the rate and the time)

   (b) time it took us to get somewhere (if we know the rate and the distance traveled)

Extensions

- Students can convert their rate of speed from feet/second to miles/hour and find the average rate of speed for the class using these new terms.

Name ______________________ Date ______________

How Fast Can You Throw?

How fast can you throw? We can find out an approximation of the speed of our throws by carrying out the following experiment:

- Each person will throw a ball against a wall three times and find the average time it takes to hit the wall.
- If everyone stands a known distance from the wall and we have an average time, we can use this formula to find the rate or speed the ball traveled:

$$\frac{\mathbf{d}}{\mathbf{t}} = \mathbf{r}$$

where **d** equals the distance from the wall, **r** equals the rate or speed, and **t** equals the average time it took to hit the wall.

Each student's rate will be placed on a class chart and a class average will be obtained.

Directions: Complete this sheet. Then place your average or mean rate on the class chart for analysis.

Your Name		
Trial	**Distance from the Wall**	**Time in Seconds**
1		
2		
3		
Average		

(continued)

Name ________________________________ Date ________________

How Fast Can You Throw? *(continued)*

Class Data Sheet

Name of Student	Average Rate of Throw in Feet/Second
Average Rate of Throw	

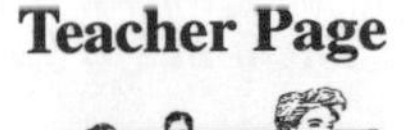

Walking the Indy 500

Areas of Study

Data collection, problem solving, measurement, averaging

Concepts

Students will:

- measure an agreed-upon distance
- use a stopwatch to measure the amount of time each member of the group takes to walk this distance
- find the average time for the group
- convert the rate of speed from feet/second to miles/hour
- use this new rate to calculate the average amount of time it would take the group to walk the Indy 500
- work cooperatively and share duties

Materials

- stopwatch (for each group of four)
- calculator (for each group)
- Walking the Indy 500 handouts (for each group)
- yardstick, tape measure, or trundle wheel (for each group)
- clipboard (for each group)

Procedure

Divide class into groups of four. Explain to students that they will work in their groups to measure their rate of speed in walking a predetermined distance. Discuss with the class how long they want this distance to be.

For example, each group may measure 100 feet and every member of the group will walk (and be timed) for this distance. Discuss whether a distance of 20 feet is appropriate. What factors might affect their ability to record accurately the time it would take to walk 20 feet? 50 feet? 100 feet?

Once the whole class has agreed upon a distance, give each group the materials it needs and assign these jobs:

- **timer**—person who uses the stopwatch to measure the time
- **recorder**—person who records the time
- **calculator**—person who calculates the averages
- **walker**—person walking the predetermined distance

Each person in the group should have a turn doing each of the jobs. When the group has calculated its average rate in feet per second, it will need to convert this ratio to miles per hour. This can be done by setting up the following ratio:

$$\frac{\textbf{Time (in seconds)}}{\textbf{Distance (in feet)}} = \frac{\textbf{3,600 seconds (\# of seconds in 1 hour)}}{\textbf{5,280 feet (\# of feet in 1 mile)}}$$

Once the group knows how many miles per hour it can average, it should find out how many hours it would take them to walk the 500 miles of the Indy 500.

Assessment

1. Observation of student
2. Student data collection sheet
3. Journal question:
   - Explain how your group converted its data from feet per second into miles per hour. What did you do next to calculate how long it would take you to walk the Indy 500?

Extensions

- Students can repeat the experiment to determine their running speed. After calculating their rate of speed in miles per hour, they can answer the question: "How long would it take you to *run* the Indy 500?"

Internet Connection

Use the Internet to research the history of the Indy 500 and other car races that are run in the U.S. Where are they run, how long are they, and who are the record holders in these races?

A good site to find out more about the Indy 500 is:

http://www./Indy500/.com

Other race-related sites:

http://www.speedworld.net/

http://www.nascar.com/

Name ______________________ Date ______________

Walking the Indy 500

The Indy 500 is a car race held every Memorial Day weekend. It is named for the town in which it is run—Indianapolis, Indiana—and its length—500 miles.

How long would it take you to walk the length of the Indy 500? To answer this question we are going to conduct an experiment.

Directions: Work with your group to:

1. measure an agreed-upon distance.
2. using a stopwatch, time how long it takes for each member of the group to walk that distance.
3. find the average amount of time it takes your group to walk the distance.
4. calculate your speed in miles per hour.
5. use this average to calculate how long it would take your group to walk 500 miles or the Indy 500.

Use the table below to record your data. Be sure to show your calculations.

Name	Distance Walked	Time	Rate of Speed $\frac{\text{Distance (in feet)}}{\text{Time (in seconds)}}$
Average			

You have now calculated your rate of speed in feet per second. Problem-solve with your group to convert this to miles per hour. Now you have all the information you need to calculate how long (on average) it would take your group to walk the Indy 500. Show your work here:

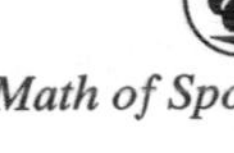

Design Your Own Game

Areas of Study

Scale drawing, area, scoring, calculation of playing time, communication, symbol design

Concepts

Students will:

- design an original game to include:
  - (a) name
  - (b) object
  - (c) number that can play
  - (d) equipment needed
  - (e) playing field and area of field
  - (f) length of play
  - (g) rules
  - (h) method to determine a winner
  - (i) method of evaluating the game
  - (j) symbol to represent the game

Materials

- access to outdoor area to plan and measure playing field
- Design Your Own Game handouts
- yardsticks, tape measures, or trundle wheels
- calculators

Procedures

Before students begin this project, discuss examples of rules and procedures for playing games. Students may give examples of games they have invented and played. Many neighborhood games have special rules that adapt to the area. Students should be asked to give examples of rules that have been modified or changed to fit local areas or playing fields. Students should also have access to an outdoor area to plan and measure the playing field, so that they can draw and calculate the area of the field. After the games have been designed and you have read the designs, time should be found to allow students to play the games.

Assessment

1. Observation of student
2. Grading rubric
3. Completion of design and worksheets
4. Journal questions:
   - In professional sports, many statistics are recorded for each player and team. What type of statistics could be collected for your game?
   - In your opinion, what was the most difficult part of designing your own game?
   - How did you determine what symbol you would use to represent your game? Why?

Extensions

- Using the library or the Internet, investigate the rules of professional sports such as the National Basketball Association, Women's National Basketball Association, National Football League, Women's Professional Indoor Volleyball Association, Professional Golfers Association, Ladies Professional Golfers Association, or amateur sports sponsored by the National College Athletic Association.
- Design a team logo and professional arena for your sport.

Name ______________________________ Date ____________________

Design Your Own Game

Grading Rubric

	Possible Points	Actual Points
1. Name of Game	1	______
2. Object of the Game • clearly stated • reasonable or attainable	2	______
3. Number of Participants	1	______
4. Equipment • clearly stated • safety of players is addressed	2	______
5. Playing Field • drawn to scale • clearly labeled • out-of-bounds areas clearly shown • safety factors are considered	4	______
6. Area • calculations are correct • area is reasonable	2	______
7. Length of Play • clearly stated • reasonable for participants • each participant will have a turn	3	______
8. Rules • clearly stated • easy to understand • fair to all participants • appropriate for participants and ability of players • method for mediation of disputes	5	______
9. Method of Determining Winner • clearly stated • fair to each team	2	______
10. Symbol • represents the sport • creatively done	2	______
Total points possible:	24	Student total: ______

(continued)

Name ______________________________ Date ________________

Design Your Own Game *(continued)*

Many of today's sports have origins in ancient games. The Olympics are based on games played over 2,000 years ago. While many games are adaptations of older games, some developed more recently. In addition, games may change to suit a particular place. For example, softball leagues adapt rules for local fields and customs. A form of softball is played on the beaches of California that defines a "home run" as a ball that is hit into the ocean. The variations of the tides, winds, and waves affect the outcome of the game.

In this exercise, you will design your own outdoor game. You may adapt the game for your area, age, abilities, and interests, as well as those of your classmates. You must think of the object of the game, length of time the game is played, number of players, number of teams, rules, playing area, equipment, scoring, and how to determine a winner (or even if you **need** a winner). What will you call your new game? The following pages will help you design your own game.

1. What is the name of your new outdoor game?

2. In baseball, the object is to score more runs than the opponent. In soccer, the object is to score more goals. In swimming, the object is to have the fastest time. What is the object of your game?

3. How many people will play your game? Is it an individual sport, or a game for two or more players? For two or more teams?

4. What kind of equipment does your game require? Be sure to include safety equipment your players will need. For example, in soccer you would need a ball, goals, and shin guards. You may also design totally new and different equipment.

(continued)

Name ______________________ Date ______________________

Design Your Own Game *(continued)*

5. Design the playing field. Draw the playing field to scale in the area below. What will be out of bounds? Are any areas off-limits because of safety? Be careful of traffic, smaller children, windows, doors, trees, and other students.

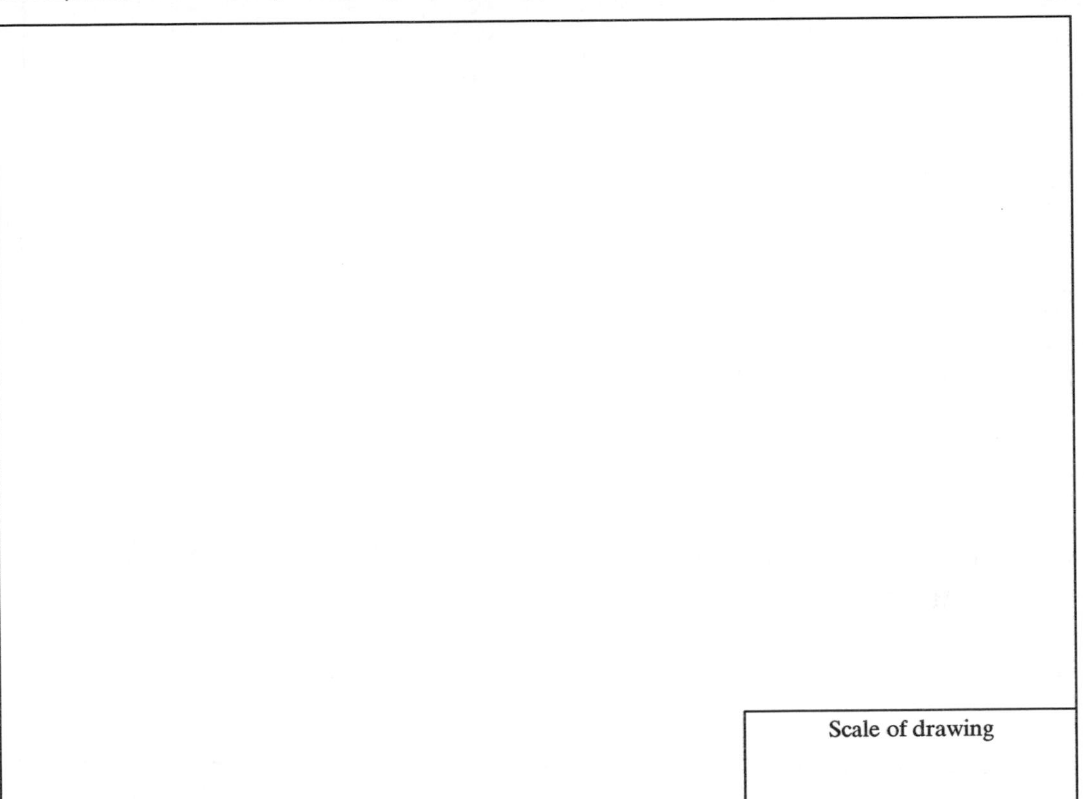

6. Figure out the area of your playing field in square feet. Show your calculations.

7. How long will you play? When will you stop? Will your game last a set time, until recess is over, or until a certain score in reached?

(continued)

Name ____________________ Date ____________________

Design Your Own Game *(continued)*

8. What are the rules of your game? For example: Who goes first? What counts as a foul? Who can call timeouts?

9. Will your game have a winner? Who wins? How do you determine who wins?

10. If possible, have some classmates play the new game you designed. Ask them to write an evaluation of the game. Did the game seem fair to all sides? Did the rules make sense? Was it easy or hard to learn? How would you make improvements?

11. Each Olympic sport has a symbol. Design a symbol to represent your new game.

Integer Football

Areas of Study

Computation, positive and negative integers, addition and subtraction of positive and negative integers

Concepts

Students will:

- use integers to determine location on a football grid
- add and subtract integers

Materials

- scissors
- small containers for game pieces
- clock, stopwatch, or timer
- copy of the football field (for every pair of students)
- set of the game cards (cut apart) and a "football" (for each pair)
- two copies of the Integer Football—Score Sheet (for each pair)

Preparation

Before play can begin, the game pieces and football must be cut out. Place the game pieces in containers that allow for easy mixing of pieces. Small margarine containers with lids or paper lunch bags work well.

Procedures

Divide the class into pairs. The object of the game is to score the most touchdowns in a set period of time (determined by the teacher). Five to ten minutes is suggested.

Each student pair receives one set of game cards, one football grid, one football, and two copies of the Integer Football—Score Sheet. Before the game begins, announce the length of each game.

Rules of the game appear on the Student Score Sheet. The black line on the football should be used to mark the location on the football grid.

Remind students that they must record the location of the football on their score sheets after each move.

Assessment

1. Observation of student during game
2. Completed worksheet
3. Journal questions:
   - How are addition and subtraction shown to be opposite operations on the football field?
   - Did the game cards seem fair? Why or why not?
   - Using the time allowed for a game and the number of touchdowns scored, find the average time needed to score a touchdown.

Extensions

- This game only allows for the scoring of touchdowns. Design a game and new rules that allow for other plays.
- The game cards allow movement in steps of ten. Design a football grid and game cards that would allow steps of one point. This game may require much more time to play.

Name ______________________ Date ______________

Integer Football—Score Sheet

The object of this game is to score as many "touchdowns" as you can during a set amount of time. You will use positive and negative numbers to move the football up and down the football field. The amount of yardage gained or lost will be determined by randomly selecting game cards. A gain of yards for you will be a loss of yards for your opponent.

The football is put into play on the 50-yard line. Each player selects a game card. The student with the highest integer value goes first, moving the football on the field as indicated on the game card. After the game card is used, it is mixed back in with the other cards.

Players take turns selecting game cards and moving the football up or down the field. Both players record the location of the football at the end of each play. When a player reaches or passes the 100-yard line, a touchdown is recorded and the football is returned to the 50-yard line. The player who did not score a touchdown now begins the play. The player with the most touchdowns when time is called wins the game.

Record the movements of your football in this area. Write "TD" each time you score a touchdown.

The winner of this game of Integer Football was ______________________

Name ______________________ Date ______________

Integer Football—Football Field

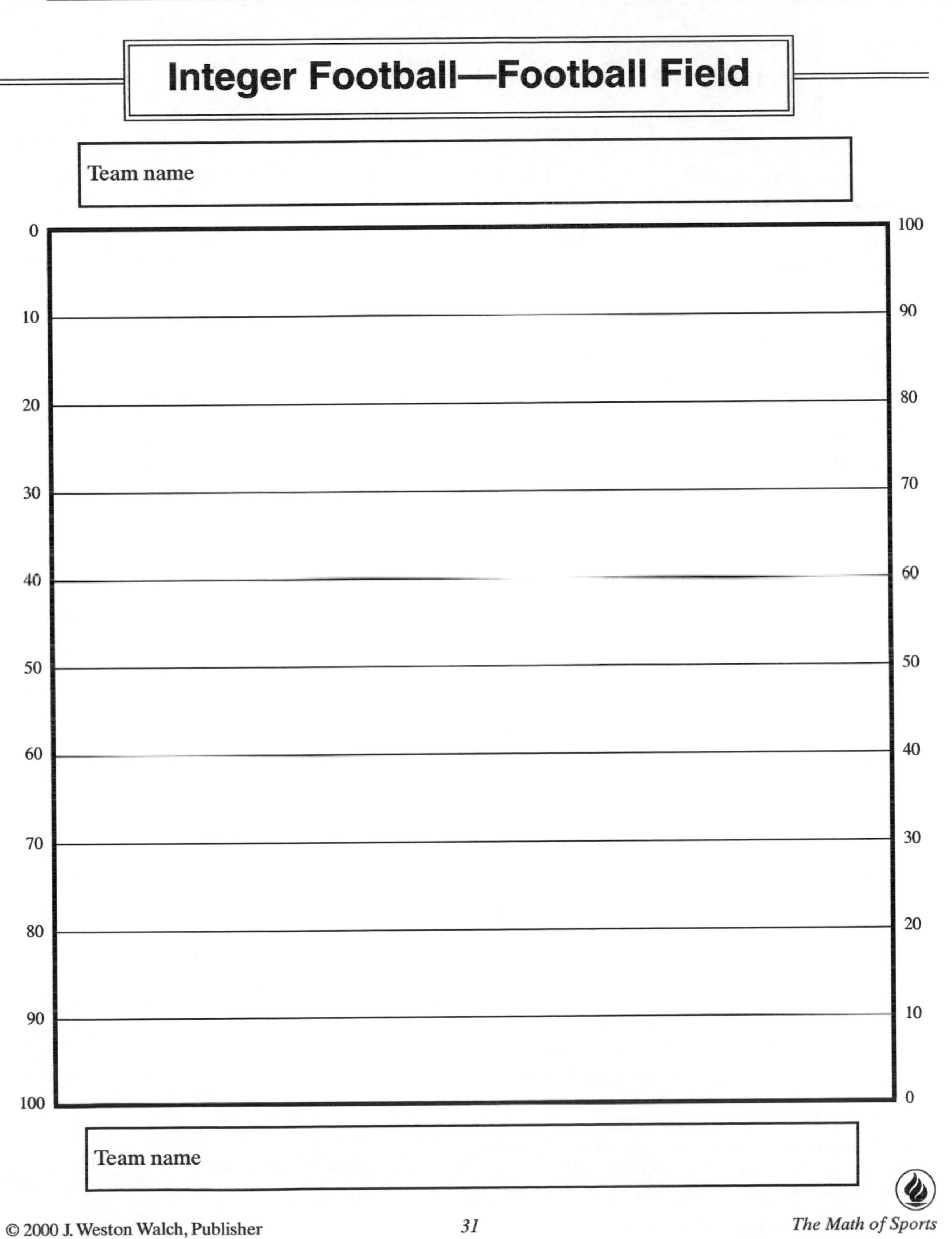

Name ______________________________ Date ______________________

Integer Football—Game Cards

Integer Football

Integer Football

Integer Football

0	0
+10	+10
+10	+10
+20	+20
–10	–10
–10	–10
–20	–20
10-yard penalty	10-yard penalty
Lost possession; Lose your turn	Lost possession; Lose your turn
Incomplete pass	Incomplete pass
0	0
Pass +30	Pass +30
Touchdown	Touchdown
–30	–30
Touchdown	Touchdown
Great play +40	Great play +40
Lost possession; Lose your turn	Lost possession; Lose your turn
Touchdown	Touchdown
Pass interception –40	Pass interception –40

Is the Super Bowl Boring?

Areas of Study

Data collection, organization, and analysis; percentages; computation; range, mean, median, and mode; graphing; stem-and-leaf and circle graphs; Roman numerals

Concepts

Students will:

- organize data from past Super Bowl games
- calculate the point spread of the final scores
- calculate the mean, mode, median, and range
- construct a stem-and-leaf graph
- categorize point spreads by touchdowns
- calculate the percentage in each category
- construct a circle graph

Materials

- Super Bowl handouts (one set per pair of students)
- calculators
- rulers
- protractors
- markers or colored pencils
- data and worksheets for each student

Procedure

1. Initiate a class discussion on the question, "What makes a football game boring?" Explain that the **point spread** is the difference between the winning and losing scores. If one team is winning by a large point spread, the game can become boring. For this exercise a point spread of more than 3 touchdowns or 21 points is considered boring.
2. Divide the class into pairs. Give each pair of students a Super Bowl data sheet, a set of Super Bowl scores and the worksheets. The Super Bowl Scores sheet includes data from 1967 to 2000. If you wish, you can have students research more recent scores and add them to the data sheet.
3. On their data sheets, students will convert Super Bowl Roman numerals into Arabic numbers and calculate the point differences.
4. Using these point differences, students should calculate the range, mean, mode, and median of the scores, and then plot a stem-and-leaf graph. They will next analyze the data and calculate what percentage of the Super Bowl games are boring.
5. Finally, students will categorize scores by point spread, calculate the percentages and degrees of a circle, and construct a circle graph.

Solutions

Year	Super Bowl Number Roman Numeral	Super Bowl Number Arabic Numbers	Winning Score	Losing Score	Point Spread (Winning–Losing)
1967	I	1	35	10	25
1968	II	2	33	14	19
1969	III	3	16	7	9
1970	IV	4	23	7	16
1971	V	5	16	13	3
1972	VI	6	24	3	21
1973	VII	7	14	7	7
1974	VIII	8	24	7	17
1975	IX	9	16	6	10
1976	X	10	21	17	4
1977	XI	11	32	14	18
1978	XII	12	27	10	17
1979	XIII	13	35	31	4
1980	XIV	14	31	19	12
1981	XV	15	27	10	17
1982	XVI	16	26	21	5
1983	XVII	17	27	17	10
1984	XVIII	18	38	9	29
1985	XIX	19	38	16	22
1986	XX	20	46	10	36
1987	XXI	21	39	20	19
1988	XXII	22	42	10	32
1989	XXIII	23	20	16	4
1990	XXIV	24	55	10	45
1991	XXV	25	20	19	1
1992	XXVI	26	37	24	13
1993	XXVII	27	52	17	35
1994	XXVIII	28	30	13	17
1995	XXIX	29	49	26	23
1996	XXX	30	27	17	10
1997	XXXI	31	35	21	14
1998	XXXII	32	31	24	7
1999	XXXIII	33	34	19	15
2000	XXXIV	34	23	16	7

1. Largest point spread: Game XXIV, 45 points
2. Smallest point spread: Game XXV, 1 point
3. Range of point spreads: 45 – 1 = 44

4. Student stem-and-leaf graphs should look like this:

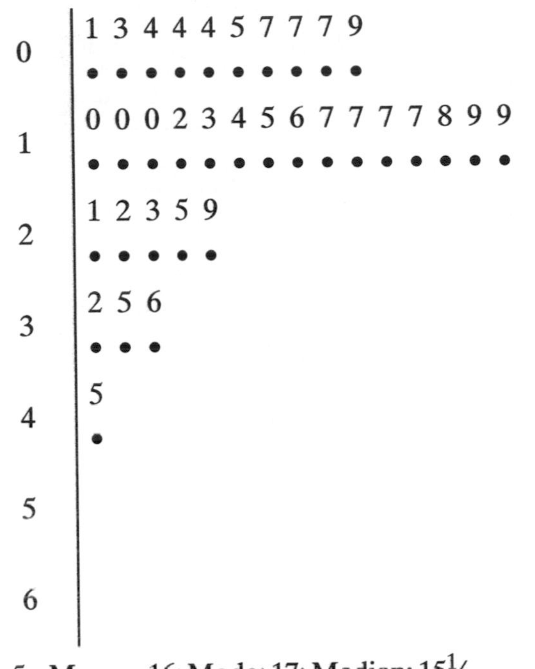

5. Mean: ~16; Mode: 17; Median: $15\frac{1}{2}$
6. Answers will vary. Student opinion.
7. Answers will vary. Student opinion.
8. If a point spread of 21 points constitutes a boring game, ~26 percent of Super Bowl games have been boring.
9. Point-spread chart and circle graph:

Point Spread	Number of Games	Percentage of Games	Degrees of Circle (to Nearest Degree)
Point spread is from 1 to 7	9	26.5	95
Point spread is from 8 to 14	7	20.6	74
Point spread is from 15 to 21	10	29.4	106
Point spread is 22 or greater	8	23.5	85
Totals	34	100	360

Student circle graphs should look like this:

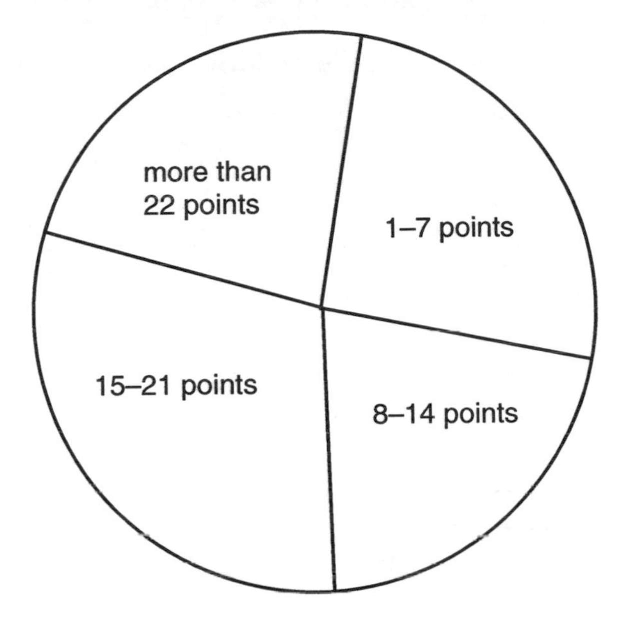

Assessment

1. Observation of student pairs
2. Completed worksheets and graphs
3. Classroom discussion of boring football games
4. Journal questions:
   - Were you surprised by the point spread data?
   - Do you agree that a point spread of more than 21 points makes a boring game?
   - Using the point spread data and the statistics you calculated, what do you predict the point spread of this year's Super Bowl will be?

Extensions

- Use statistics and graphs to compare the point spreads of the first five Super Bowls and the last five years shown on the table. Are Super Bowls getting more boring?

Name ______________________________ Date ____________________

Is the Super Bowl Boring? Scores

Directions: Complete the data sheet by writing the number of each Super Bowl in Arabic numbers and calculating the point spread for each game.

Year	Super Bowl Number Roman Numeral	Super Bowl Number Arabic Numbers	Winning Score	Losing Score	Point Spread (Winning–Losing)
1967	I	1	35	10	25
1968	II	2	33	14	19
1969	III		16	7	
1970	IV		23	7	
1971	V		16	13	
1972	VI		24	3	
1973	VII		14	7	
1974	VIII		24	7	
1975	IX		16	6	
1976	X		21	17	
1977	XI		32	14	
1978	XII		27	10	
1979	XIII		35	31	
1980	XIV		31	19	
1981	XV		27	10	
1982	XVI		26	21	
1983	XVII		27	17	
1984	XVIII		38	9	
1985	XIX		38	16	
1986	XX		46	10	
1987	XXI		39	20	
1988	XXII		42	10	
1989	XXIII		20	16	
1990	XXIV		55	10	
1991	XXV		20	19	
1992	XXVI		37	24	
1993	XXVII		52	17	
1994	XXVIII		30	13	
1995	XXIX		49	26	
1996	XXX		27	17	
1997	XXXI		35	21	
1998	XXXII		31	24	
1999	XXXIII		34	19	
2000	XXXIV		23	16	

Name ______________________________ Date ________________

Is the Super Bowl Boring?

Directions: Use the Super Bowl Scores sheet to complete this page.

1. Which Super Bowl game had the largest point spread? ________________
2. Which game had the smallest point spread? ________________
3. Calculate the range of the point spread. ________________
4. Using the point-spread data from the chart, sort point difference by the tens digit on the first stem. Using the sorted data, make a stem-and-leaf graph on the second stem.

	0	
	1	
	2	
	3	
	4	
	5	
	6	

5. Calculate the following statistics from the point-spread data.

   Mean: ________________

   Mode: ________________

   Median: ________________

6. What do you consider a boring football game? In your opinion, how large a point spread makes a football game boring? ________________
7. If a team is 3 touchdowns (21 points) ahead, would the game become boring? ________________
8. For this math exercise, consider a point spread of 21 points or more as a boring game. What percent of the Super Bowl games have been boring? ________________

(continued)

Name ______________________________ Date ________________

Is the Super Bowl Boring? *(continued)*

9. Use the point spread data to complete the chart. Then, use the point-spread data and the chart to construct a circle graph. Label the graph.

Point Spread	Number of Games	Percentage of Games	Degrees of Circle (to Nearest Degree)
Point spread is from 1 to 7 Lost by 1 touchdown or less			
Point spread is from 8 to 14 Lost by more than 1 touchdown but no more than 2			
Point spread is from 15 to 21 Lost by more than 2 touchdowns but no more than 3			
Point spread is 22 or greater Lost by more than 3 touchdowns			
Totals			

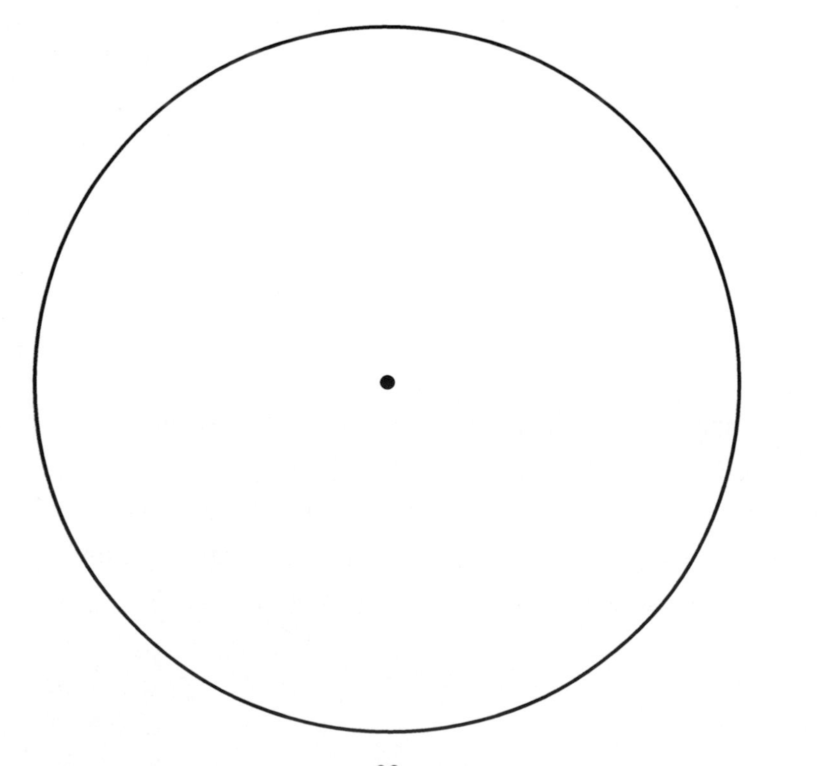

At the Olympic Table

Areas of Study

Open-ended problem solving, computation, unit cost, volume

Concepts

Students will:

- research the unit cost of foods
- find the total cost for the amount of food used at the summer Olympics
- calculate the cost of food per athlete, per day, and per participating country
- find the volume of a rectangular prism
- estimate the amount of space rectangular prisms would occupy
- estimate the capacity of a normal bathtub

Materials

- newspaper ads or information from a local supermarket
- calculators
- At the Olympic Table handouts (one per pair of students)

Procedure

The first activity contains a shopping list that gives the amount of food used at a summer Olympics. Students are asked to find the cost per unit of each item. For example, 1.2 million pounds of beef and lamb were used. At an average cost of $2.00 per pound, this would cost $2,400,000. Students are asked to research the cost of the items and find the total cost of each of the items on the list. They must then calculate the cost/athlete, cost/participating country, and the cost/day for the food.

The second activity has a diagram of a box of pasta with the dimensions on the box. Students must calculate the volume of one box and find how many cubic inches (in.^{3}) of space 52,000 boxes would take up.

The third activity is a homework assignment. Students are asked to calculate the volume of their bathtubs. Measurements should be made to the very top of the tub. While size will vary, the average capacity is between 40 and 50 gallons. However, variations should be discussed and explored.

Solutions

1. Answers will vary depending on the unit price identified by students.
2. Answers will vary depending on the unit price identified by students.
3. (a) 495,000
   (b) Answers will vary depending on the unit price identified by students.
4. The volume of each box is $11 \times 9 \times 2$ or 198 in.^{3} Therefore, 52,000 boxes would be $198 \times 52{,}000$ or 10,296,000 in^3. To convert this volume to appropriate measurements, students need to find the number of cubic feet (ft^3) this represents: $12" \times 12" \times 12" = 1{,}728$ ft^3 $10{,}296{,}000 \div 1{,}728 \approx 5{,}958$ ft^3 of space.
5. Answers will vary. If your classroom is 40 ft $\times$ 40 ft $\times$ 8 ft, it has a volume of 12,800 ft^3.
6. Answers will vary, but if a bathtub has a capacity of 45 gallons, the athletes drank about 1,555.5 "bathtubs" full of milk.

Assessment

1. Observation of student
2. Completed worksheets
3. Journal questions:
   - Explain the procedures you used to convert cubic inches to cubic feet.

- If one egg has 80 calories, how many calories would there be in 48,000 dozen? Explain your answer.

Extensions

- Students can attempt to estimate the amount of space other items on the list might take up, or how many calories would be contained in others. For example, one tablespoon of butter contains 100 calories and there are 32 tablespoons in one pound! How many calories would there be in 30,000 pounds of butter?

Internet Connection

An interesting site to explore both the ancient and modern Olympics is:

http://olympics.tufts.edu/

Here you will find sections about the ancient and modern Olympic games, a tour of ancient Olympia, the context of the games and the Olympic spirit, and the athletes' stories.

Name ______________________________ Date ______________

At the Olympic Table

How much food does it take to feed nearly 11,000 competitors from 197 countries? In addition, there will be about 4,000 coaches and other officials to feed. How much would all this food cost? The table below gives some indication of the enormous quantities of food on our shopping list.

Work with your partner to calculate the approximate cost of this shopping expedition. Use newspaper ads or visit a local market to get the unit prices you need.

Food	Quantity	Cost per Unit	Total Cost for Item
Asparagus	15,498 lb		
Tomatoes	17,998 lb		
Beef	600,000 lb		
Lamb	600,000 lb		
Poultry	750,000 lb		
Strawberries	23,342 pt		
Shredded Cheese	9,057 lb		
Eggs	48,000 doz		
Radicchio	30,000 lb		
Coconut Custard Pies	2,656		
Rice	34,000 lb		
Pasta	52,000 lb		
Milk	70,000 gal		
Apples	750,000		
Butter	30,000 lb		
Peaches	226,000		
Bean Sprouts	2,800 lb		
Tabouli	2,500 lb		
Garlic	1,034 lb		
Bottled Water	550,000 gal		
TOTAL COST OF SHOPPING SPREE			

1. If there are 11,000 hungry athletes, how much did the shopping spree cost per athlete? ________
2. If each of the 197 countries were assessed an equal share of the cost, how much would each country spend to feed the athletes? ________

(continued)

Name ______________________________ Date ______________________

At the Olympic Table *(continued)*

3. How many millions of meals will be served over 33 days? __________ How much will feeding the athletes cost per day? ______________________

4. Think of a box of ziti (a type of pasta). The box holds 1 lb of pasta; its dimensions are 11" × 9" × 2". How much space would 52,000 lbs, or 52,000 boxes, fill? Show your work in the space provided.

Use this space to describe how you calculated the volume of 52,000 boxes of pasta. Be sure to convert your answer into appropriate units of measure. ______________________

____

____

____

5. What kind of container do you think would hold this much pasta? Would it fit in your classroom? ____

____

6. The athletes and their coaches drank 70,000 gallons of milk. About how many gallons of water do you think your bathtub holds? Would you be surprised to learn that it is probably less than 50 gallons?

   At home, calculate the capacity or volume of your bathtub. Then compute the number of bathtubs full of milk the athletes consumed. Use the space below to describe how you calculated the number of full bathtubs.

____

____

____

____

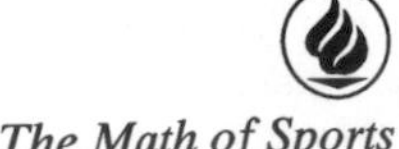

Cost of Television Coverage of the Summer Olympics

Areas of Study

Computation, percent increase, bar graph, analysis of data

Concepts

Students will:

- calculate the difference of the cost of U.S. coverage of the Olympic games
- calculate the percentage increase in cost from the previous Olympic games
- make a bar graph to show the cost of television coverage, the increase in the cost of television coverage, and the percent increase in the cost of coverage.

Materials

- calculators
- Cost of Television Coverage handouts for each student
- rulers

Procedure

The cost of television coverage of sports events has risen sharply as the income potential of television advertising profit has increased. The cost of U.S. coverage of the Olympics in 2004 will be about 200 times the cost of coverage of the 1960 Olympics. In this exercise, students will calculate the increase in cost and the percentage increase in cost, and make bar graphs to show the relationship of these data. The percentage increase is found by using the following formula:

$$\frac{\textbf{Increased Dollars from Previous Olympics}}{\textbf{Cost of Distribution of Previous Olympics}} = \frac{\%}{100}$$

Solutions

Year	Cost	Increase in Dollars	Percentage Increase
1964	$ 1,500,000	1,106,000	281%
1968	$ 4,500,000	3,000,000	200%
1972	$ 7,000,000	2,500,000	56%
1976	$ 10,000,000	3,000,000	43%
1980	$ 87,000,000	77,000,000	770%
1984	$225,000,000	138,000,000	159%
1988	$300,000,000	75,000,000	33%
1992	$401,000,000	101,000,000	34%
1996	$456,000,000	55,000,000	14%
2000	$705,000,000	249,000,000	55%
2004	$793,000,000	88,000,000	12%

1. Based on the Cost of Television Coverage bar graph, students should predict that the cost to televise the Olympics in the U.S. will be over one billion dollars in the year 2012.
2. Years of greatest increase in cost: 1964 (281%), 1968 (200%), 1980 (770%), and 1984 (159%). Students may notice that some of the most drastic increases came in years when the network handling the coverage changed.

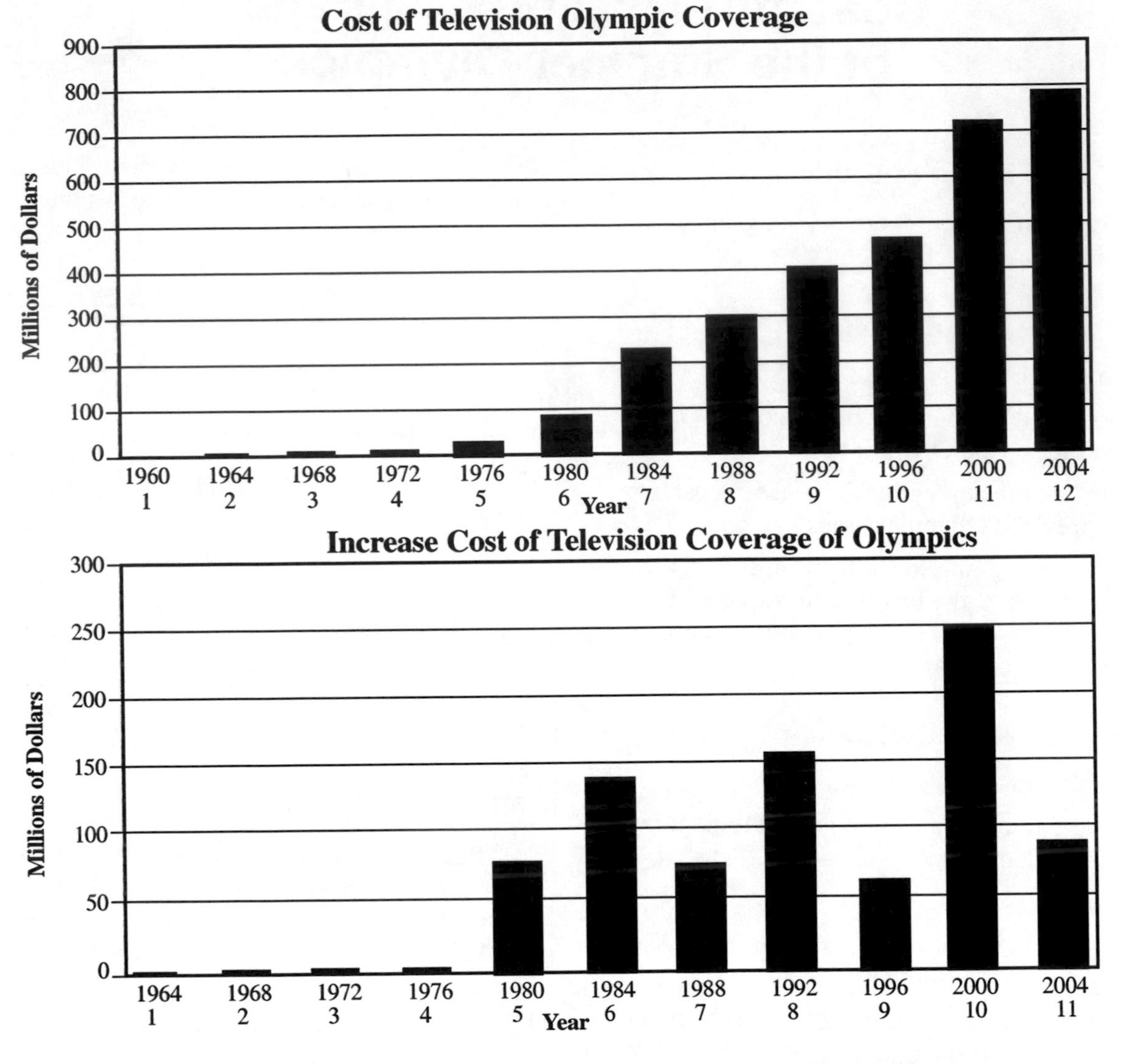

Assessment

1. Observation of student work
2. Completed worksheet and bar graphs
3. Journal questions:
   - Coverage of the 1960 Olympics cost $394,000. If the cost doubled every four years, what would be the projected cost to televise the 2016 games?
   - In 1964, the dollar increase in the cost of coverage was small but the percentage increase was the second largest. How can this happen mathematically?
   - What effect does competition have on the price or value of any object?

Extensions

- Find the average cost of U.S. coverage of the Olympic games. What is the range of cost? What other statistical values can you calculate from this data?
- Investigate the cost of a one-minute advertisement during an important sporting event.
- During one hour of a televised sporting event, use a stopwatch to record the length and type of advertisements. Then, determine the fraction and percent of each hour of programming used for ads. Classify the ads by product or age group of the potential customer and make a graph to show what you discovered.

Name ______________________________ Date ______________

Cost of Television Coverage of the Summer Olympics

The cost of bringing the summer Olympics to the American public on television has increased since coverage first began. Complete the chart below to find the dollar amount of increase from the previous Olympic games. Then determine the percentage increase from the previous games. Round the percentage increase to the nearest whole number. Use the formula:

$$\frac{\textbf{Increase in Dollars from Previous Olympics}}{\textbf{Cost of Distribution of Previous Olympics}} = \frac{\%}{100}$$

Year	Network	Cost for U.S. Distribution	Increase in Dollars from Previous Olympics	Percentage Increase from Previous Olympics
1960	CBS	$ 394,000		
1964	NBC	$ 1,500,000		
1968	ABC	$ 4,500,000		
1972	ABC	$ 7,000,000		
1976	ABC	$ 10,000,000		
1980	NBC	$ 87,000,000		
1984	ABC	$225,000,000		
1988	NBC	$300,000,000		
1992	NBC	$401,000,000		
1996	NBC	$456,000,000		
2000	NBC	$705,000,000		
2004	NBC	$793,000,000		

Use the completed chart above to make the three bar graphs that follow. One shows the cost of U.S. distribution of the Olympic games. Another shows the increased cost of bringing the summer Olympic games to the U.S. The third bar graphs shows the percent increase in the cost in television coverage of the Olympics.

(continued)

Name ______________________________ Date ______________

Cost of Television Coverage of the Summer Olympics *(continued)*

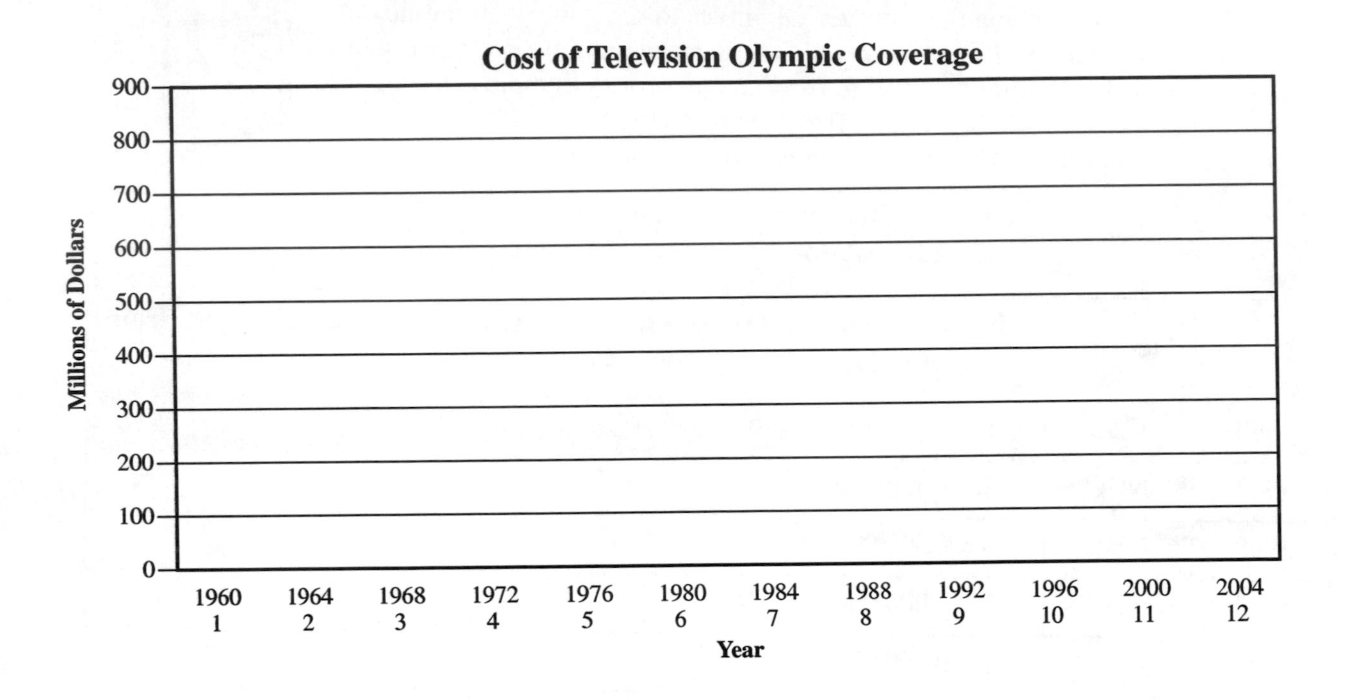

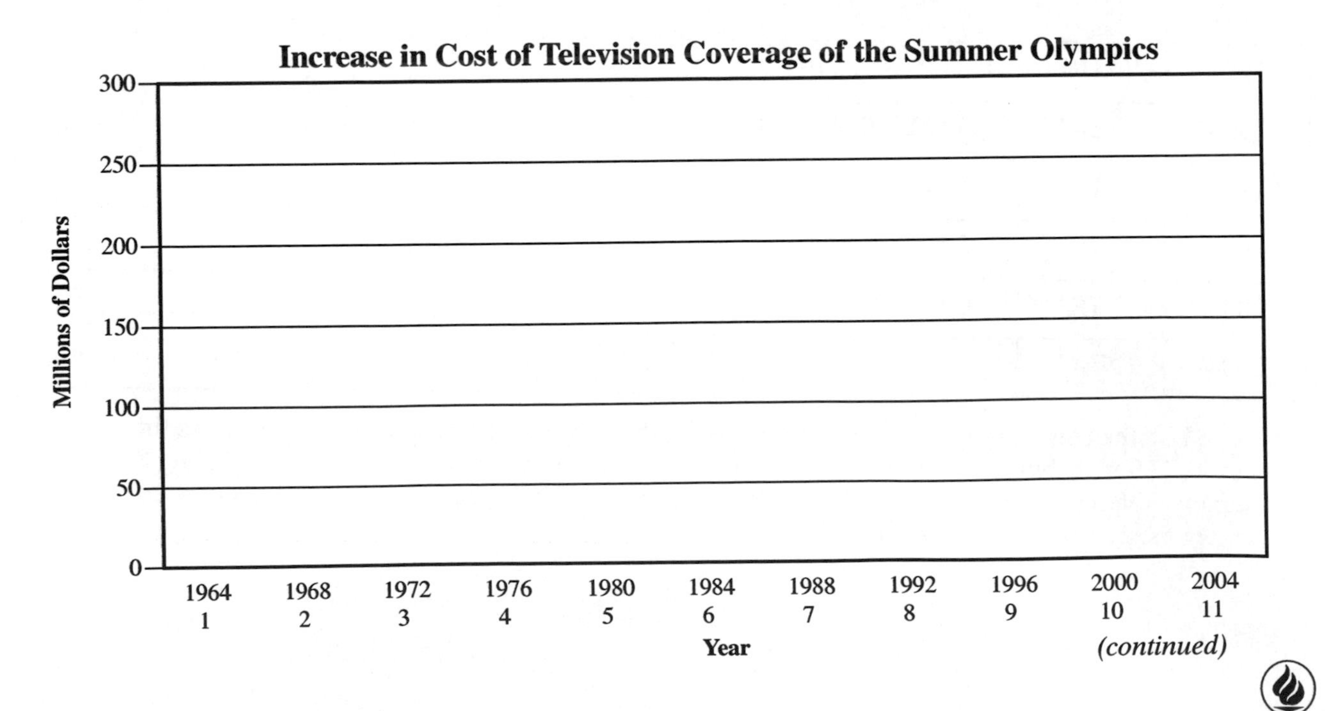

(continued)

Name ______________________________ Date ______________

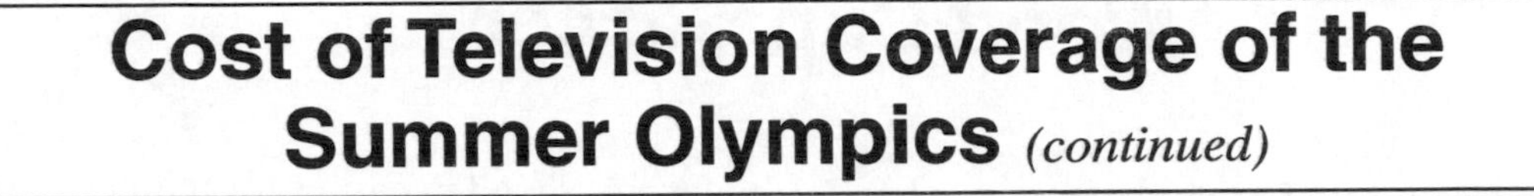

Cost of Television Coverage of the Summer Olympics *(continued)*

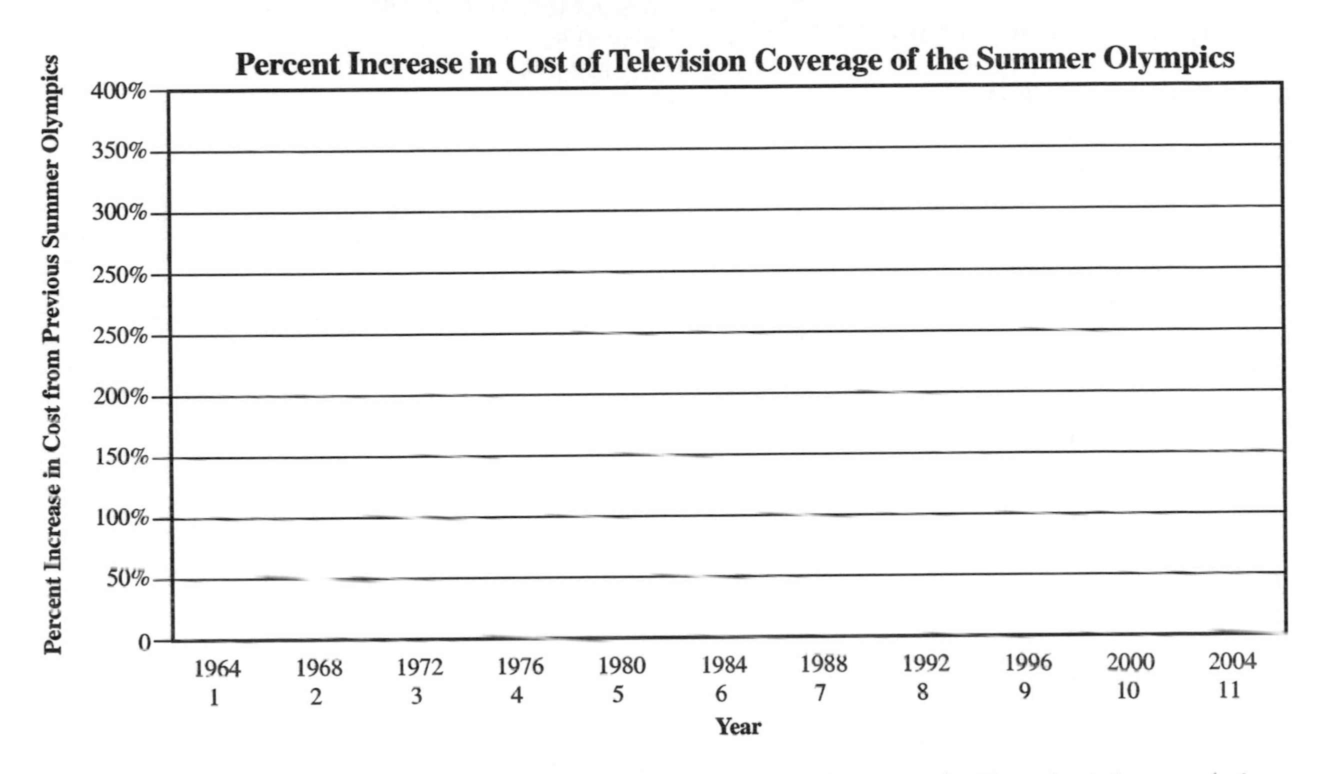

1. Look at the Cost of Television Coverage bar graph you have made. Based on the graph, in what year do you predict that the cost to televise the Olympics in the U.S. will be over one billion dollars? ____________

2. Examine the graph that shows the percent increase in cost. In what years was the increase the greatest? ____________ Why do you suppose the percent increase was greatest in those years? ______________________________

____

3. In 1998, the Fédération Internationale de Football Association (FIFA), the international soccer association, was criticized for selling world rights to televise the World Cup for 10 million dollars. In your opinion, could FIFA have received more money? ____________

____

World Cup Soccer

Areas of Study

Calculation, percent, average, percents, ratios

Concepts

Students will:

- calculate the average attendance of the World Cup games
- calculate the percentage of games won by the host country
- calculate the percentage of games won by South American teams
- predict the attendance of a future World Cup

Materials

- World Cup Soccer Attendance handouts
- calculators

Procedures

Before beginning, ask students what they know about the World Cup soccer tournament. The Fédération Internationale de Football Association (FIFA) has sponsored the tournament, which is held every four years, since 1930. Cumulatively, 37 billion people watched the 1998 World Cup on television. If needed, data for 2002 and 2006 may be obtained from an almanac or from the Internet.

The average attendance is calculated by dividing the total attendance by the number of games played. The average attendance of the 1950 World Cup in Brazil was not surpassed until the 1994 tournament. After calculating average attendance, students are asked to calculate the percentage of games won by Brazil, percentage of games hosted by a South American country, percentage of games won by South American teams, and percentage of World Cup tournaments that had an average attendance over 50,000. More games are played now than were played in earlier tournaments. To predict the total attendance if 52 games were played in the 1950 World Cup instead of 22, use the following formula:

$$\frac{22}{52} = \frac{1{,}337{,}000}{X}$$

Solutions

Soccer Attendance	
Year	**Average Attendance**
1930	24,111
1934	23,235
1938	26,833
1950	60,773
1954	36,269
1958	24,800
1962	24,250
1966	50,459
1970	50,312
1974	46,685
1978	42,374
1982	33,967
1986	42,307
1990	48,282
1994	68,604
1998	43,366

1. 4; 25%
2. 25% (Mexico is in North America); 50%
3. 37.5%
4. 25%
5. 3,160,000

World Cup Trophy

Value of gold in the World Cup trophy:

$11 \times 16 \times \$300 \times 75\% = \$39{,}600$

Assessment

1. Observation of student
2. Completed worksheet
3. Classroom discussion of World Cup and soccer
4. Journal questions:
   - Using the data on the chart, write another question involving percentages for your classmates to solve.
   - When did the largest increase in attendance from the preceding World Cup occur?
   - In what years did the attendance decrease?

Extensions

- Use the given data to make a bar-and-circle graph showing the number of times each country has won the World Cup.
- Investigate the attendance of other major sporting events such as the Olympics, World Series, and Super Bowl. Compare these data to World Cup attendance.

Internet Connection

The Fédération Internationale de Football Association can be found at:
http://www.fifa.com/

Name ____________________ Date ____________________

World Cup Soccer Attendance

World Cup soccer has always drawn large numbers of enthusiastic fans. World Cup soccer was started in 1930, with the first games played in Uruguay.

The chart below shows attendance figures for the World Cup by year and the number of games played. Calculate the average number of people who attended each game.

Year	Site	Winning Team	Games Played	Total Attendance	Average Attendance
1930	Uruguay	Uruguay	18	434,000	
1934	Italy	Italy	17	395,000	
1938	France	Italy	18	483,000	
1950	Brazil	Uruguay	22	1,337,000	
1954	Switzerland	West Germany	26	943,000	
1958	Sweden	Brazil	35	868,000	
1962	Chile	Brazil	32	776,000	
1966	England	England	32	1,614,677	
1970	Mexico	Brazil	32	1,673,975	
1974	West Germany	West Germany	38	1,774,022	
1978	Argentina	Argentina	38	1,610,215	
1982	Spain	Italy	52	1,766,277	
1986	Mexico	Argentina	52	2,199,941	
1990	Italy	West Germany	52	2,510,686	
1994	United States	Brazil	52	3,567,415	
1998	France	France	64	2,775,400	
2002	Japan/Korea				
2006	Germany				

1. How many times has Brazil won the World Cup? __________ What percentage of the World Cup championships has Brazil won? __________
2. What percentage of World Cup tournaments have been held in South America? __________ What percentage of the World Cup tournaments have South American soccer teams won? __________
3. What percentage of World Cup games have been won by the host country? __________
4. What percent of the World Cup games had an average attendance over 50,000? __________
5. In 1950, Brazil hosted the World Cup. 22 games were played and 1,337,000 fans attended. At the same attendance rate, how many fans could be expected if 52 games were played? __________

Name ______________________________ Date ______________

World Cup Soccer Trophy

The World Cup trophy has had a turbulent history. The first World Cup trophy, made in 1930, was known as the Jules Rimet Cup, in honor of the Fédération Internationale de Football Association (FIFA) president who initiated the World Cup tournament. The trophy was gold-plated sterling silver on a base of lapis lazuli, a semiprecious stone. It was a traveling trophy, kept by the winning team until the next World Cup. During World War II, the trophy was hidden under a bed in Italy. In 1966, it was stolen from a public exhibit in England prior to the World Cup tournament. It was recovered with the help of Scotland Yard and a police dog named Pickles. The trophy was given permanently to Brazil, a three-time winner of the World Cup, in 1970. In 1983, the trophy was stolen again, from Rio de Janeiro. This time it was not recovered; most people assume that it was melted down and sold for its precious metals.

The present-day trophy was made in 1974. The FIFA has permanent possession of the trophy, and a gold-plated replica is given to the World Cup winner. The trophy is 14 inches tall, weighs 11 pounds, and is made of 18-carat gold. The trophy is $^{18}/_{24}$—that is, 75%—pure gold. If an ounce of pure gold (24 carat) costs about $300, what is the value of the gold in the World Cup trophy? Show your calculations in the space below.

Prizes in a Cereal Box

Areas of Study

Probability, data collection and analysis, averages, problem solving

Concepts

Students will:

- predict the number of boxes of cereal they must purchase to obtain one of each of the sports cards
- conduct an experiment to determine the average number of boxes needed to obtain one of each of the sports cards
- analyze their results and the class results to make a prediction of the probability of each occurrence

Materials

- Prizes in a Cereal Box handout and copy of Class Record sheet (for each student pair)
- overhead transparency of Class Record
- dice (one die for each student pair)
- calculators

Procedure

Divide the class into pairs. Discuss the accomplishments of each of the women named on the handout, using the information provided below.

Discuss the need to develop an experiment to predict the number of boxes one might need to buy in order to get one of each card. Ask, "What is the least number of boxes one would need to buy? Is there a maximum number of boxes? Why or why not?"

This experiment is a simulation that gives students the opportunity to use multiple trials (experiments) to estimate a theoretical probability from this ratio:

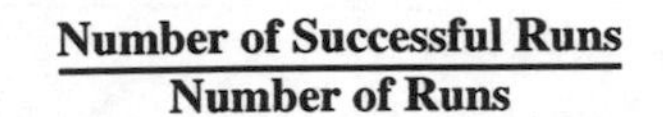

$$\frac{\text{Number of Successful Runs}}{\text{Number of Runs}}$$

When all student pairs have found the average of three trials, record the averages on the Class Record overhead, then have students find the class average.

Additional information about each of the women is provided below:

Tenley Albright

Tenley Albright was born in Boston, Massachusetts, in 1935. At the age of 11, her skating career was threatened by an attack of polio. Albright recovered, going on to win her first U.S. skating championship in 1952 and the Olympic gold medal in 1956. She retired from skating to attend Harvard University Medical School, and, in 1961, became a surgeon. She is an inductee of the Ice Skating Hall of Fame, the U.S. Figure Skating Hall of Fame, and the Olympic Hall of Fame.

Wilma Rudolf

Born in St. Bethlehem, Tennessee, in 1940, Rudolf overcame severe physical disabilities to become a championship runner. At four, she was attacked by double pneumonia and scarlet fever. The illnesses left her without the use of her left leg. With the help of physical therapy administered by her family, she began to walk at the age of eight, but could do so only if she wore a special shoe. By the age of eleven, she no longer needed the special shoe. In high school, she was an outstanding athlete. She was the first woman to win three gold medals at a single Olympics (1960) and the first woman to win both Olympic sprint events (100 m and 200 m dashes).

Annie Smith Peck

In 1850, Annie Smith Peck, a teacher, was born in Providence, Rhode Island. She first saw the Matterhorn mountain in Switzerland during a trip she took during her summer vacation. It was at that moment she decided to try mountain climbing. Her first success came in 1888, when she climbed Mount Shasta in California (14,380 feet). In 1895, she climbed the Matterhorn (14,780 feet), and then two mountains in Mexico: Mount Popocateptl (17,887 feet) and Mount Orizaba (18, 314 feet). Smith Peck was the first person to climb the north peak of Mount Huascarán, Peru (21,812 feet). Throughout her life, she continued to climb mountains and teach. The last mountain she climbed, Mount Madison in New Hampshire, is only 5,380 feet high—but Smith Peck was eighty-two years old when she made this climb.

Jackie Joyner-Kersee

Born in 1962, Joyner-Kersee was named after the wife of President John F. Kennedy because, in the words of her grandmother: "Someday this girl will be the First Lady of something." After finishing second in the 1984 Olympic Heptathlon, she won all nine events she entered during the next four years. In 1986 she became the first woman to break the 7,000-point mark with a score of 7,148 points. In 1992, just four days after winning her second Olympic Heptathlon medal, she won a second gold in the long jump.

Bonnie Blair

From Champaign, Illinois, Bonnie Blair is the youngest of six children. She has won more Olympic medals than any other American athlete. A speed skater, she won five gold medals and one bronze medal. She won the 500-meter race in 1988, 1992, and 1994. In 1992 and 1994 she won the 1,000-meter race. Her nickname "Blur" is what you see when you watch her skate.

When Bonnie Blair retired from skating in 1985, her mother wondered, "What does she do for an encore?" She needn't have worried—Bonnie's legend lives on!

Babe Didrikson Zaharias

Babe Didrikson Zaharias, born in Port Arthur, Texas, in 1914, was voted the Outstanding Woman Athlete of the Century in an Associated Press poll taken in 1950. In high school, she was the star of the girls' basketball team, and excelled in many other sports, as well. In the 1932 Olympics, she qualified for eight different events. She won two gold medals (for the javelin and 80-meter hurdles), and one silver medal (for the high jump). She took up golf in 1934 and by 1945 had won the Western Open three times. In 1947, she won seventeen straight tournaments, including the British Women's Amateur.

Assessment

1. Observation of student
2. Grading matrix
3. Journal question:
   - Explain how the probability of getting one of each card was developed in this experiment. Is there a theoretical probability that can be calculated in another way? Explain your answer.

Extensions

- Have students devise an experiment to predict the number of boxes one would need to buy if there were eight different sports cards.

Name ______________________________ Date ______________

Prizes in a Cereal Box

Morning Grains, the cereal company, is putting cards of famous U.S. women athletes from the past in their cereal boxes. The athletes are:

Tenley Albright, an Olympic figure skating champion.

Wilma Rudolph overcame polio, double pneumonia, and scarlet fever to win three Olympic gold medals in sprinting events in 1960.

Annie Smith Peck, the first person to climb the north peak of Mount Huascarán, Peru in 1908 (21,812 ft).

Jackie Joyner-Kersee, an Olympian who won nine gold medals, including the heptathlon in 1988 and 1992.

Bonnie Blair, an Olympic speed skater who has won more medals than any other U.S. athlete (five gold and one bronze).

Babe Didrikson Zaharias excelled in running, swimming, diving, high-jumping, baseball baseball, basketball, javelin, and hurdles, but her best sport was golf. She won two Olympic gold medals in 1932 (in javelin and hurdles), and one silver medal (for the high jump).

How many boxes of cereal do you think you will need to buy in order to have at least one of each of the six athletes? Write your prediction here: ____________

To find the answer we can conduct an experiment with a die.

Directions: First, we will assign a number to each athlete.

1—Tenley Albright
2—Wilma Rudolph
3—Annie Smith Peck
4—Jackie Joyner-Kersee
5—Bonnie Blair
6—Babe Didrikson Zaharias

Roll the die until each number appears at least once. Use the table below to record the number of throws you made to roll each number. Keep going until you have at least one of each of the players. Conduct the experiment three times, then find the average number of throws you needed to make to roll each number once. This is about the number of cereal boxes you would need to buy to be sure of getting one of each card.

Athlete	Trial 1	Trial 2	Trial 3
Tenley Albright (1)			
Wilma Rudolph (2)			
Annie Smith Peck (3)			
Jackie Joyner-Kersee (4)			
Bonnie Blair (5)			
Babe Didrikson Zaharias (6)			
TOTAL			

AVERAGE OF THREE TRIALS: ______________

(continued)

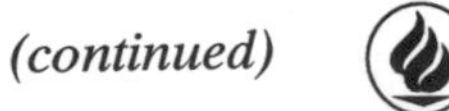

Name ______________________________ Date ______________

Prizes in a Cereal Box *(continued)*

Class Record

Group	Average Number of Trials
1	
2	
3	
4	
5	
6	
7	
8	
9	
10	
11	
12	
13	
14	
15	
Class Average	

Women in the Summer Olympics

Areas of Study

Calculation, rounding to the nearest tenth, totals, bar graphs, percents, percent increase, analysis of data

Concepts

Students will:

- calculate the total number of Olympic competitors
- calculate the percentage of women participants from the total
- calculate the percentage increase from the previous Olympic games
- construct a scatterplot showing the number of women participants in the Olympics by year
- construct a bar graph showing the percentage of women participants in the Olympics by year

Materials

- Women in the Summer Olympics handouts
- calculators

Procedure

Before beginning, discuss the ancient Olympics and the revival of the Olympic games in 1896. Women did not participate in the first Olympic games. Since 1936, the number of women participants and the percentage of women participants have risen each year.

In this activity, students are given the number of women and men participants and asked to find the total number of participants in each Olympics and the percentage of participants who were women. The percent can be found by solving the following proportion:

$$\frac{\%}{100} = \frac{\text{Number of Women Participants}}{\text{Total}}$$

Once they have completed the data table, students will use the data to construct a scatterplot showing the increase of number of women participants in the Olympic games and a bar graph showing the percent of women participants.

Solutions

Year	Location	Men	Women	Total	% Female	Increase
1896	Athens	311	0	311	0.0%	
1900	Paris	1,318	11	1,329	0.8%	0.8%
1904	St. Louis	681	6	687	0.9%	0.1%
1906	Athens	877	7	884	0.8%	–0.1%
1908	London	1,999	36	2,035	1.8%	1.0%
1912	Stockholm	2,490	57	2,547	2.2%	0.4%
1916	Canceled					
1920	Antwerp	2,543	64	2,607	2.5%	0.3%
1924	Paris	2,956	136	3,092	4.4%	1.9%
1928	Amsterdam	2,724	290	3,014	9.6%	5.2%
1932	Los Angeles	1,281	127	1,408	9.0%	–0.6%
1936	Berlin	3,738	328	4,066	8.1%	–0.9%
1940	Canceled					
1944	Canceled					
1948	London	3,714	385	4,099	9.4%	1.3%
1952	Helsinki	4,407	518	4,925	10.5%	1.1%
1956	Melbourne	2,958	384	3,342	11.5%	1.0%
1960	Rome	4,738	610	5,348	11.4%	–0.1%
1964	Tokyo	4,457	683	5,140	13.3%	1.9%
1968	Mexico City	4,750	781	5,531	14.1%	0.8%
1972	Munich	5,848	1,299	7,147	18.2%	4.1%
1976	Montreal	4,834	1,251	6,085	20.6%	2.4%
1980	Moscow	4,265	1,088	5,353	20.3%	–0.3%
1984	Los Angeles	5,458	1,620	7,078	22.9%	2.6%
1988	Seoul	6,983	2,438	9,421	25.9%	3.0%
1992	Barcelona	7,555	3,008	10,563	28.5%	2.6%
1996	Atlanta	7,060	3,684	10,744	34.3%	5.8%

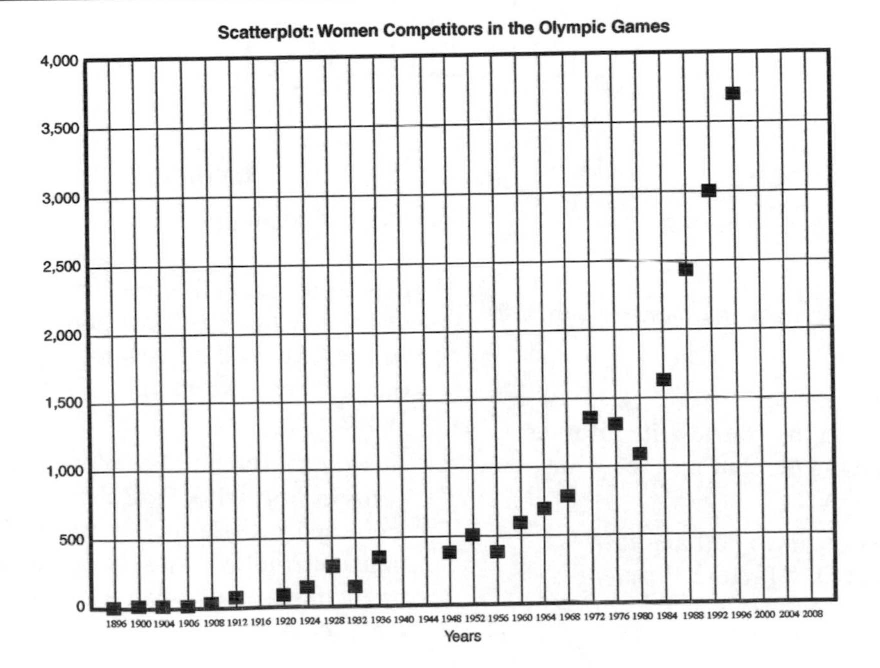

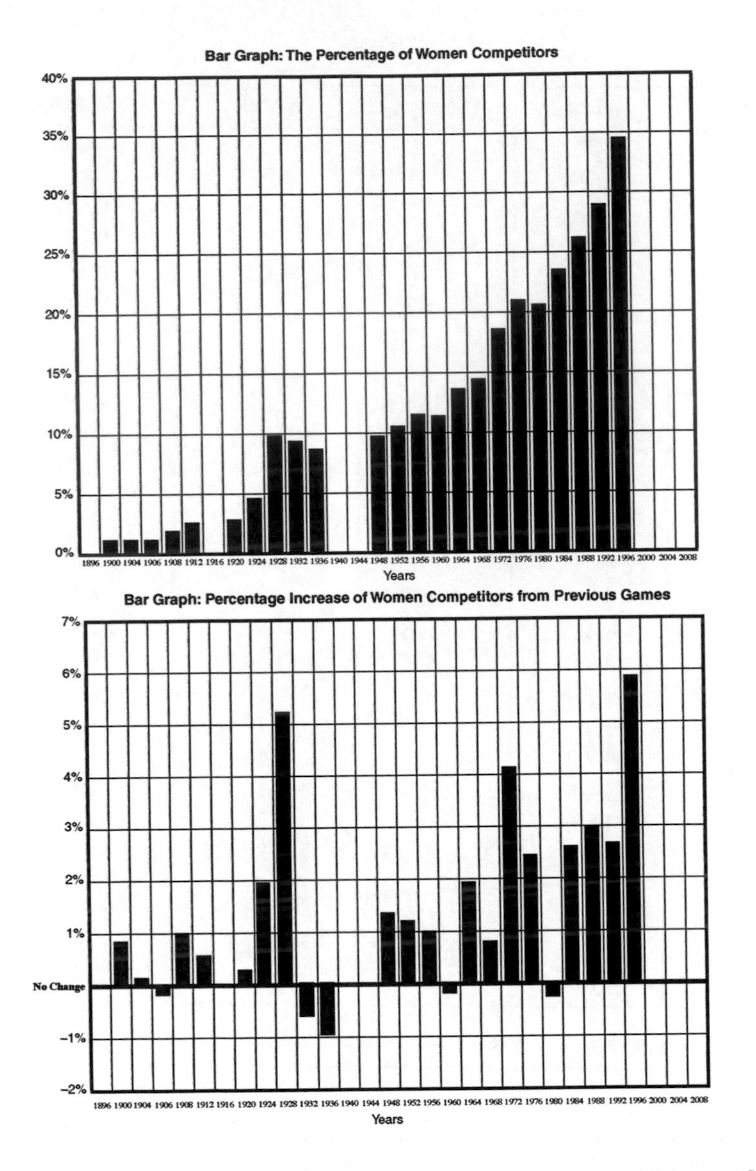

Assessment

1. Observation of student's work
2. Completed data table and graphs
3. Classroom discussion of women participants and Olympic games
4. Journal questions:
   - Using the data on the chart, write another question involving percentages for your classmates to solve.
   - What other sports have had a great increase in women athletes? Why?

Extensions

- Use the given data to make another bar graph showing the increases in the percentages of women athletes from the preceding Olympic games. Some of the increased percentages will be negative, even though the number of women participants increased. Why?
- Investigate the number of women participants in your school's athletic program. Has it increased in the last 20 years? 10 years? What influence did federal school funding have on the number of women participants in the athletic program?

Name ______________________ Date ______________

Women in the Summer Olympics

The Summer Olympic Games began in 1896 and have been held every four years since then, except for 1916, 1940, and 1944, when the games were canceled because of war. The first women participated in the 1900 games in Paris. Use the chart below to calculate the percentage of women competitors in each Summer Olympic Games. Round the answer to the nearest tenth of a percent.

When you have completed the table, use the information to create a scatterplot of women in the Summer Olympic Games, a bar graph showing the percentage of women competitors, and a bar graph showing the percentage increase from one Olympics to the next.

Year	Location	Male Competitors	Female Competitors	Total	Percent Female	Increase
1896	Athens	311	0			
1900	Paris	1,318	11			
1904	St. Louis	681	6			
1906*	Athens	877	7			
1908	London	1,999	36			
1912	Stockholm	2,490	57			
1916	Canceled					
1920	Antwerp	2,543	64			
1924	Paris	2,956	136			
1928	Amsterdam	2,724	290			
1932	Los Angeles	1,281	127			
1936	Berlin	3,738	328			
1940	Canceled					
1944	Canceled					
1948	London	3,714	385			
1952	Helsinki	4,407	518			
1956	Melbourne	2,958	384			
1960	Rome	4,738	610			
1964	Tokyo	4,457	683			
1968	Mexico City	4,750	781			
1972	Munich	5,848	1,299			
1976	Montreal	4,834	1,251			
1980	Moscow	4,265	1,088			
1984	Los Angeles	5,458	1,620			
1988	Seoul	6,983	2,438			
1992	Barcelona	7,555	3,008			
1996	Atlanta	7,060	3,684			
2000	Sydney					
2004	Athens					

*1906 games were held as a ten-year anniversary of the modern Olympics.

(continued)

Name ______________________ Date ______________

Women in the Summer Olympics *(continued)*

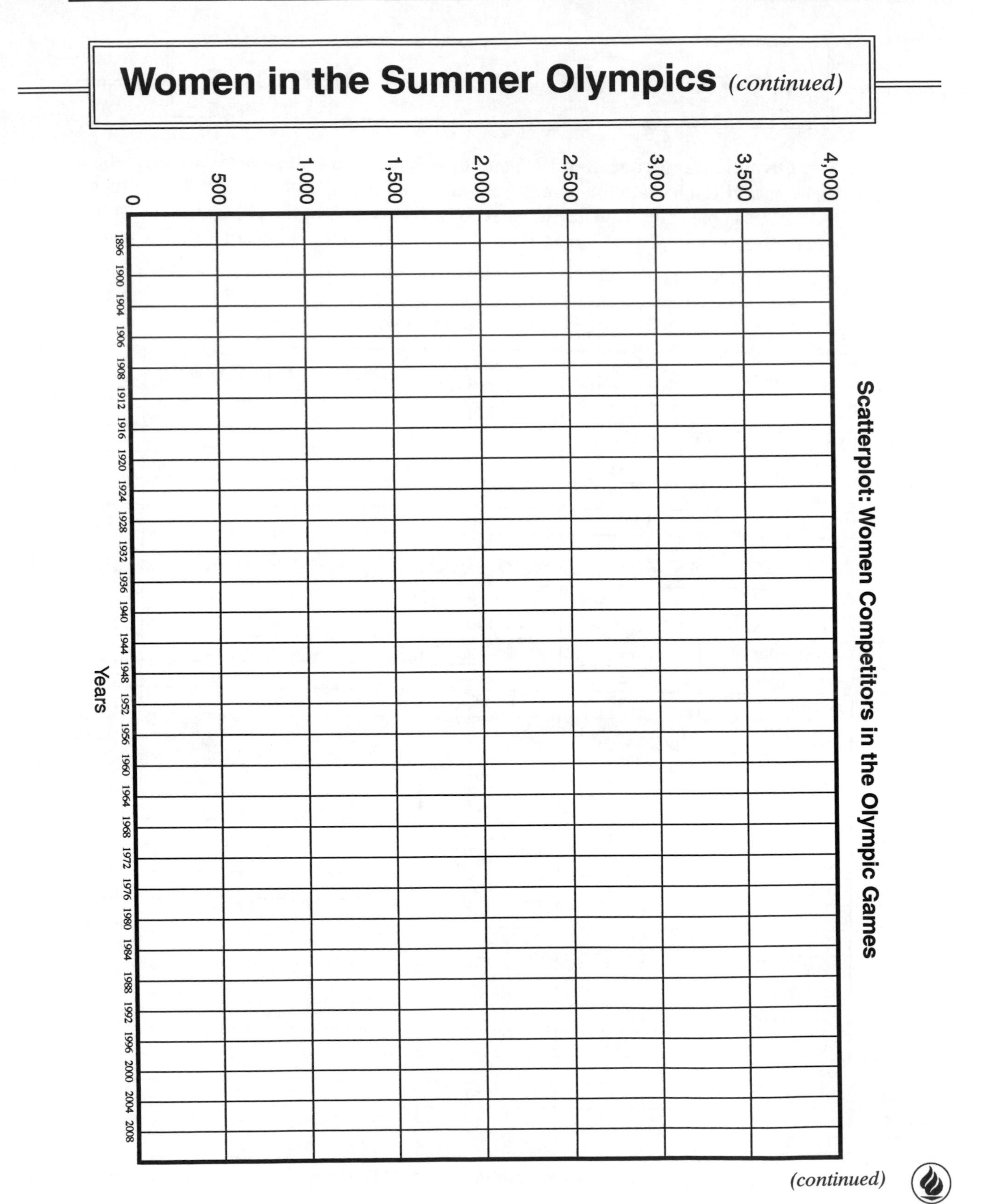

(continued)

Name ______________________ Date ______________

Women in the Summer Olympics *(continued)*

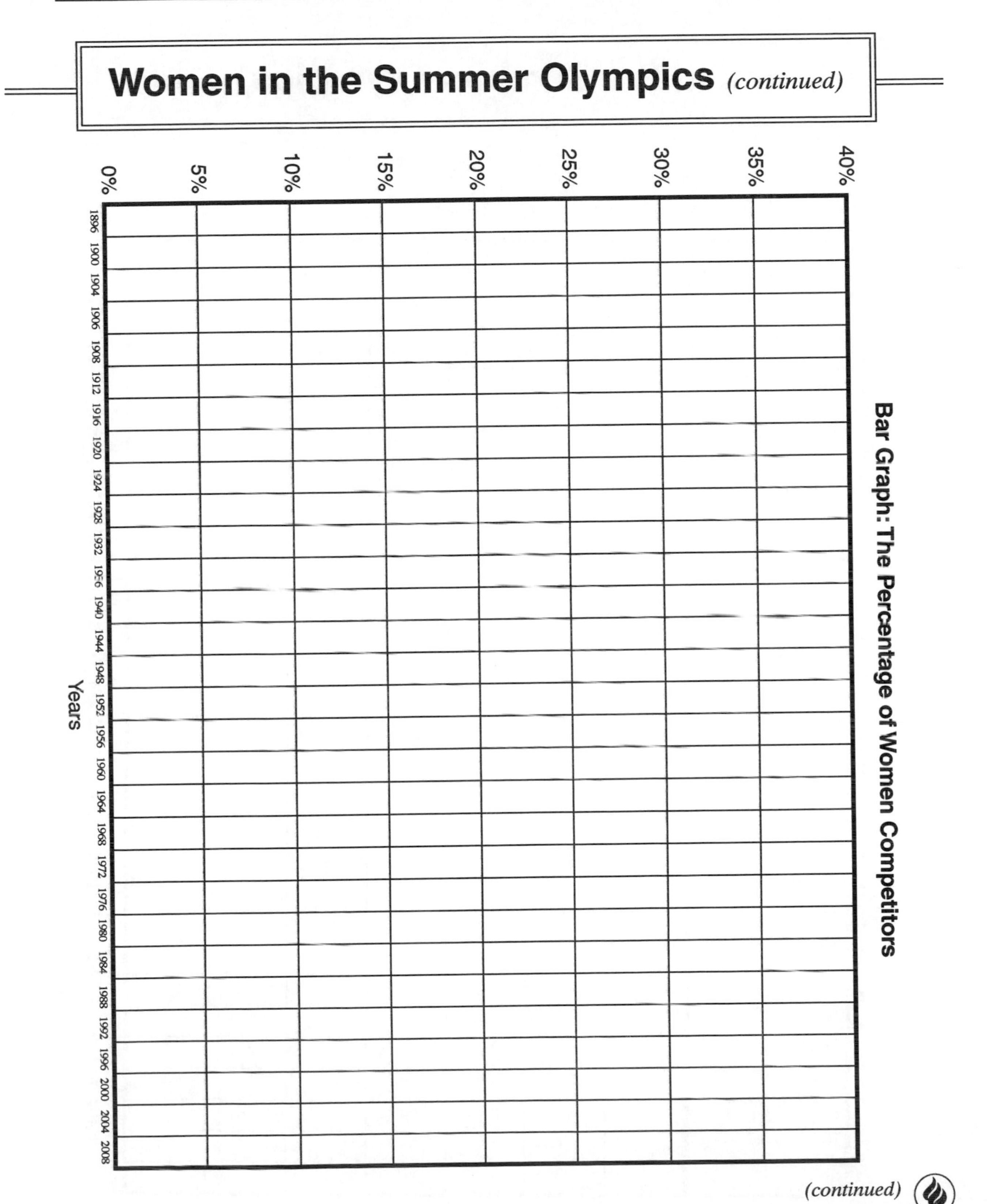

(continued)

Name ____________________ Date ____________________

Women in the Summer Olympics *(continued)*

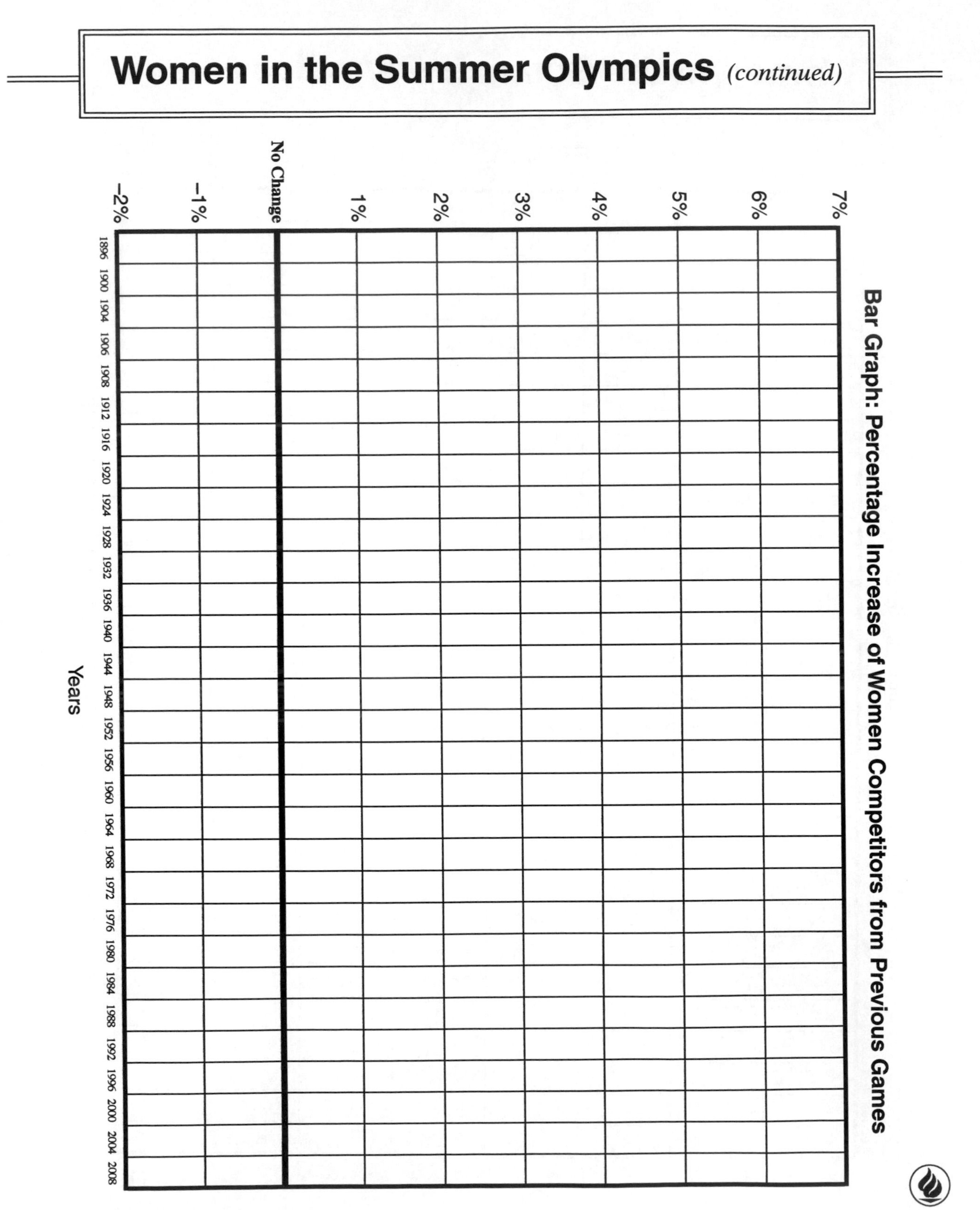

Running the Iditarod

Areas of Study

Scale drawing, ratio, proportion, computation, percentages, reading tables and charts, problem solving, measurement conversions

Concepts

Students will:

- Use a map scale to calculate distances on a map
- Convert time differences into days, hours, minutes, and seconds
- Calculate the percentage of increase in prize money
- Use Iditarod data to analyze interesting facts

Materials

- Running the Iditarod handouts
- calculators (optional)
- string (to help students measure winding distances)
- rulers

Procedure

Discuss the historical beginnings of the Iditarod, which has become the most famous and popular of dog sledding races. The map of Alaska shows three different trails for the race: the historical path (by which the serum was delivered in 1925), the Northern Route, and the Southern Route.

In the first part of the activity, students will measure the length of the Iditarod on a map of Alaska, using string to follow the curves of the route. Be sure students understand how to measure distances and use the map scale.

Then, they use a table of winning mushers to calculate the difference between the fastest and slowest times. Students may need to review how to borrow when they subtract in bases other than base 10.

Solutions

1. Difference between the fastest and slowest times: 11 days, 12 hours, 19 minutes, 48 seconds
2. 708%

Length of Individual Legs of Iditarod

Checkpoint	Distance in Miles	Checkpoint	Distance in Miles
Anchorage to Eagle River	20	Iditarod to Shageluk	65
Eagle River to Wasilla	29	Shageluk to Anvik	25
Wasilla to Knik	14	Anvik to Grayling	18
Knik to Yentna	52	Grayling to Eagle Island	60
Yentna to Skwentna	34	Eagle Island to Kaltag	70
Skwentna to Finger Lake	45	Kaltag to Unalakleet	90
Finger Lake to Rainy Pass	30	Unalakleet to Shaktoolik	40
Rainy Pass to Rohn	48	Shaktoolik to Koyuk	58
Rohn to Nikolai	93	Koyuk to Elim	48
Nikolai to McGrath	48	Elim to Golovin	28
McGrath to Takotna	23	Golovin to White Mountain	18
Takotna to Ophir	38	White Mountain to Safety	55
Ophir to Iditarod	90	Safety to Nome	22

Interesting Iditarod Facts

1. (a) Approximately 37.5 miles/day.
   (b) Approximately 1.6 miles/hour over 24 hours
2. About 19%
3. About 44.7 miles
4. 28 hours

Assessment

1. Observation of student
2. Grading matrix
3. Journal question:
   - How can the vast differences in times for the Iditarod racers be explained? Name all the things you think could be responsible for these time variations.

Extensions

- The Iditarod is run every March. While the race is being held, each student in class can be assigned a racer and can keep track of his or her position and times on the Iditarod web site.

Internet Connections

The official web site for the Iditarod is:
http://www.iditarod.com

It is possible to obtain teacher's packets from the committee by contacting them on this site or by writing to:

Iditarod Trail Committee
Dept. SF
1801 Parks Hwy. Ste. E 10
Wasilla, Alaska 99654-7373
(800) 545-6874

Name ______________________________ Date ______________________

Running the Iditarod

In the winter of 1925, a diphtheria epidemic threatened the inhabitants of Nome, Alaska. At that time, Nome was a remote village on the Bering Sea Coast. In winter, with temperatures around –40°F, ships could not sail to Nome. There were no roads, and the nearest railway was hundreds of miles away. Relays of dogteams raced from Nenana to Nome carrying serum to inoculate residents against the disease.

The Iditarod race was created in 1973 to honor these brave mushers. This race is a contest that challenges men, women, and dogs. The table below shows some historical data that will help you analyze this famous race:

Year	Winning Musher	Days	Hours/Minutes/Seconds	Winnings
1999	Doug Swingley	9	14:31:07	
1998	Jeff King	9	05:52:26	$51,000
1997	Martin Buser	9	08:30:45	$50,000
1996	Jeff King	9	02:42:19	$50,000
1995	Doug Swingley	9	05:43:13	$52,500
1994	Martin Buser	10	13:02:39	$50,000
1993	Jeff King	10	15:38:15	$50,000
1992	Martin Buser	10	19:17:15	$51,600
1991	Rick Swenson	12	16:34:39	$50,000
1990	Susan Butcher	11	01:53:23	$50,000
1989	Joe Runyan	11	05:24:34	$50,000
1988	Susan Butcher	11	11:41:40	$30,000
1987	Susan Butcher	11	02:05:13	$50,000
1986	Susan Butcher	11	15:06:00	$50,000
1985	Libby Riddles	18	00:20:17	$50,000
1984	Dean Osmar	12	15:07:33	$24,000
1983	Rick Mackey	12	14:04:44	$24,000
1982	Rick Swenson	16	04:40:10	$24,000
1981	Rick Swenson	12	08:45:02	$24,000
1980	Joe May	14	07:11:51	$12,000
1979	Rick Swenson	15	10:37:47	$12,000
1978	Dick Mackey	14	18:52:24	$12,000
1977	Rick Swenson	16	16:27:13	$ 9,600
1976	Jerry Riley	18	22:58:17	$ 7,200
1975	Emmitt Peters	14	14:43:15	$15,000
1974	Carl Huntington	20	15:02:07	$12,000
1973	Dick Wilmarth	20	00:49:41	$12,000

1. Use the information on the table to calculate the difference between the fastest and slowest times recorded by the winners of the Iditarod. Express your answer in days, hours, minutes, and seconds. ______________________________
2. The first-place winner in 1976 won $7,200; the first-place winner in 1998 won $51,000. What is the percentage of increase in these two purses? ______________________

(continued)

Name ______________________________ Date ______________

Running the Iditarod *(continued)*

This map shows the route of the Iditarod in both even and odd years. Notice that the race always starts in Anchorage and ends in Nome. But when the racers get to the checkpoint at Ophir, racers take the Northern Route in even number years. In odd number years, they take the Southern Route. Both routes meet back at Kattag.

Notice the scale located in the bottom-left-hand corner of the map. Use this scale to approximate the length of each section of this famous dog race during an odd number year. Use these checkpoints to help you keep track of your measurements. The total distance of the Southern Route is 1,161 miles!

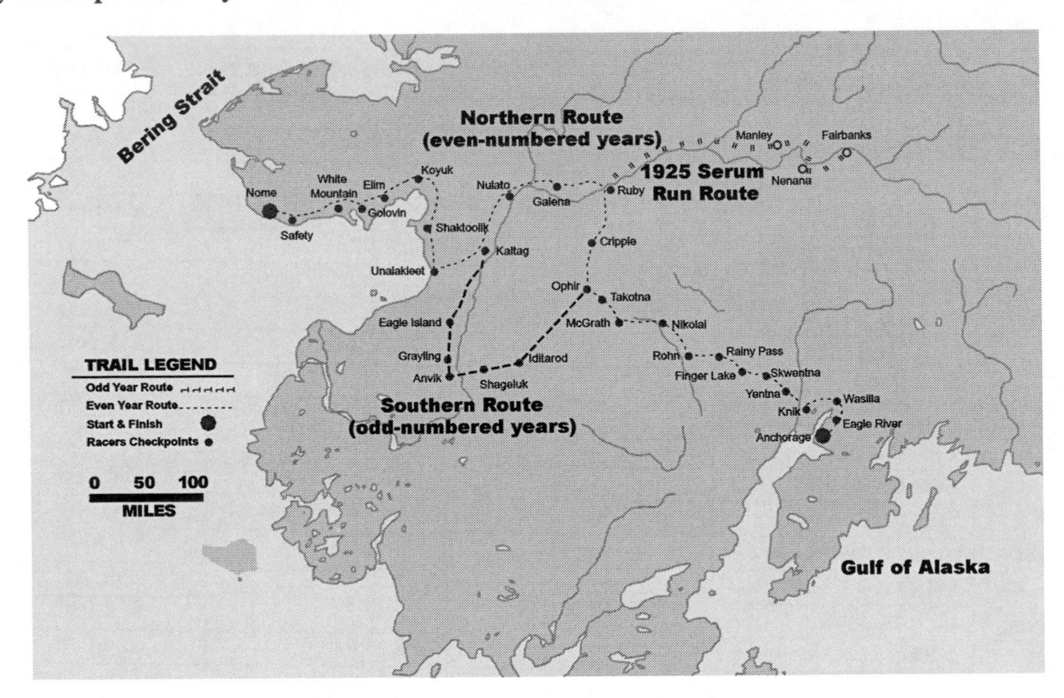

Length of Individual Legs of Iditarod

Checkpoint	Distance in Miles	Checkpoint	Distance in Miles
Anchorage to Eagle River	________	Iditarod to Shageluk	________
Eagle River to Wasilla	________	Shageluk to Anvik	________
Wasilla to Knik	________	Anvik to Grayling	________
Knik to Yentna	________	Grayling to Eagle Island	________
Yentna to Skwentna	________	Eagle Island to Kaltag	________
Skwentna to Finger Lake	________	Kaltag to Unalakleet	________
Finger Lake to Rainy Pass	________	Unalakleet to Shaktoolik	________
Rainy Pass to Rohn	________	Shaktoolik to Koyuk	________
Rohn to Nikolai	________	Koyuk to Elim	________
Nikolai to McGrath	________	Elim to Golovin	________
McGrath to Takotna	________	Golovin to White Mountain	________
Takotna to Ophir	________	White Mountain to Safety	________
Ophir to Iditarod	________	Safety to Nome	________

Name ______________________ Date ______________

Some Interesting Iditarod Facts

The first woman to finish the Iditarod was Mary Shields. She raced in 1974 (the second year of the race). Another woman, Lolly Medley finished a little while later in the day. It took Mary about 28 days to run the race. If the race is 1,049 miles long, how many miles did she average each day? What was her average rate of speed (in miles per hour)?

____

Rick Swenson has won the race five times (more times than any other person). Between 1973 and 1998 there were 26 races. What percentage of those races did Swenson win?

____

The Southern Route of the Iditarod is 1,161 miles long. There are 26 checkpoints. What is the average distance between checkpoints?

____

The distance between the Eagle Island and Kaltag checkpoints on the Southern Route is 70 miles. If a sled dog team travels at an average speed of 2.5 miles per hour, how long will it take for the team to get from Eagle Island to Kaltag?

____

Web Sites for the Mathematics of Sports

Mathematics education sources are helpful sites for educators. The sites below provide excellent resources.

The Mathematics Forum
http://forum.swarthmore.edu/

Eisenhower National Resources Clearinghouse for Mathematics and Science Education
http://www.enc.org/rf/nf_index.htm

MacTutor History of Mathematics
http://www-history.mcs.st-and.ac.uk/history/

MSTE (a database of mathematics lessons using the Internet)
http://www.mste.uiuc.edu/tcd/curriculum/matrix.html

Education World
http://db.education-world.com/perl/browse?cat_id=1515

MathMania
http://www.theory.csc.uvic.ca/~mmania/index.html

Missouri Middle School Math Project
http://www.coe.missouri.edu/~mathed/M3/

National Council of Teachers of Mathematics
http://www.nctm.org/

PBS Teachers Source
http://www.pbs.org/teachersource/math.htm?default

Math Archive
http://archives.math.utk.edu/

Professional Sports Teams

National Basketball Association—from this site you can access all NBA teams and the latest news
http://www.nba.com/

Women's National Basketball Association—from this site you can access all WNBA teams and the latest league and player news
http://www.wnba.com/index.html

National Football League—from this site you can access all NFL teams and the latest news
http://www.nfl.com/

Professional Golfers Association—lists the PGA schedule, news, and golfers statistics
http://www.pga.com/

World Cup Soccer

http://www.soccernet.com/euro2000

http://www.fifa.com/

Olympic Games

The United States Olympic Committee
http://www.olympic-usa.org/

The International Olympic Committee
http://www.olympic.org/help/plugins.html

The IOC has a list of sports organizations and web sites.
http://www.olympic.org/ioc/e/org/if/list_all_e.html

Baseball

http://www.majorleaguebaseball.com/

http://baseball-almanac.com/

http://www.blackbaseball.com/

http://www.dlcwest.com/~smudge/index.html

Car Racing

National Association for Stock Car Automobile Racing
http://www.nascar.com/

Indianapolis Race Way
http://www.irace.com/

Sports Scores

The Sporting News
http://www.sportingnews.com/

CBS Sports Line
http://cbs.sportsline.com/

ESPN Sports Zone
http://espn.go.com/

USA Today Sports News
http://www.usatoday.com/sports/sfront.htm

NCAA

Complete site of all NCAA schedules for all sports, list of college participants, and scholarship information
http://www.ncaa.org/

Glossary of Terms and Formulas

area of circle πr^2

area of rectangle length × width ($A = lw$)

area of trapezoid $\frac{1}{2}(b_1 + b_2)h$

area of triangle $\frac{1}{2}bh$

box-and-whisker plot A type of graph that uses the medians of a set of data. (1) First find the median or middle of the data. Half the data is now on either side of this point. (2) Find the median of the lower half—this is called the lower quartile. (3) Find the median of the upper half—this is called the upper quartile. (4) After constructing a number line, place a dot at the lowest data point and a dot at the highest data point; connect these two points to form a line. (This is the whisker.) The whisker represents the range of the data. (5) A box is drawn at the lower quartile, median, and upper quartile. Fifty percent of the data falls within this box. Twenty-five percent of the data is below and twenty-five percent is above. Example of a box-and-whisker plot:

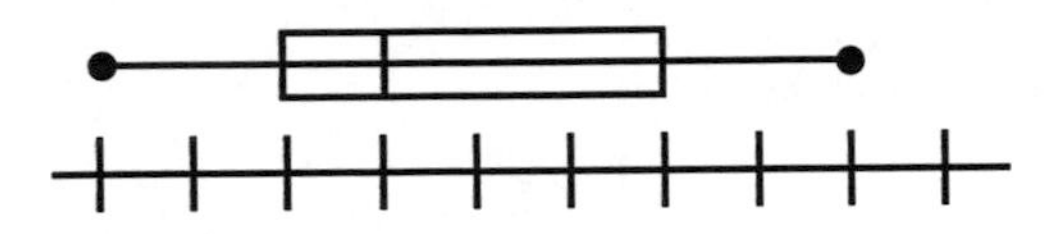

circumference of a circle πd or $2\pi r$

frequency The count of the number of pieces of data

line plot A graphic representation. Individual pieces of data are indicated on a number line by using an "x" to stand for each event. This type of graph shows clumps of data. Notice in the example below the data that are clumped around 4 and 5:

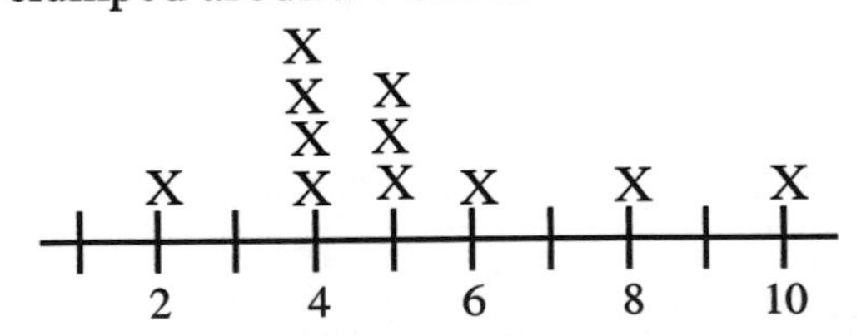

mean The arithmetic average determined by finding the sum of the data and dividing that sum by the number of pieces of data

median The middle number when a set of numbers are arranged in order of size. Fifty percent of the data is less than the median; fifty percent of the data is greater than the median.

mode A statistical term indicating the piece of data or score that appears the most often. There can be more than one mode.

net	A two-dimensional representation of a three-dimensional polyhedron
perimeter of rectangle	$2(1 + w)$ or $2w + 21$
polyhedron	A three-dimensional figure whose faces are comprised of polygons
prime number	A number with two and only two factors, 1 and itself
Pythagorean theorem	Formula developed to find the length of a missing side of a right triangle. The sum of the squares of the legs is equal to the square of the hypotenuse: $a^2 + b^2 = c^2$
range	A statistical term found by finding the difference between the largest and smallest pieces of data
scatterplot	A graphic representation that shows the relationship between two different sets of data and can be used to predict trends (positive, negative, or no correlation between the data)
stem-and-leaf plot	A style of graphic representation. The *stem* of the graph contains the tens place and the *leaf* of the graph contains the digits in the ones place, e.g.: | 4 | 3 means 43
surface area of a sphere	$4\pi r^2$
volume of polyhedra	Area of the base × height
volume of sphere	$\frac{4}{3}\pi^3$

We want to hear from you! Your valuable comments and suggestions will help us meet your current and future classroom needs.

Your name____Date____________________

School name________________________________Phone__________________________

School address____

Grade level taught_______Subject area(s) taught_______________________Average class size_____

Where did you purchase this publication?____

Was your salesperson knowledgeable about this product? Yes_____ No_____

What monies were used to purchase this product?

___School supplemental budget ___Federal/state funding ___Personal

Please "grade" this Walch publication according to the following criteria:

Quality of service you received when purchasing .... A B C D F
Ease of use .... A B C D F
Quality of content .... A B C D F
Page layout .... A B C D F
Organization of material .... A B C D F
Suitability for grade level .... A B C D F
Instructional value .... A B C D F

COMMENTS:____

What specific supplemental materials would help you meet your current—or future—instructional needs?

Have you used other Walch publications? If so, which ones?_______________________________

May we use your comments in upcoming communications? ___Yes ___No

Please **FAX** this completed form to **207-772-3105**, or mail it to:

Product Development, J.Weston Walch, Publisher, P.O. Box 658, Portland, ME 04104-0658

We will send you a **FREE GIFT** as our way of thanking you for your feedback. **THANK YOU!**